Optimization of Urban Wastewater Systems using Model Based Design and Control

DISSERTATION

Submitted in fulfilment of the requirements of
the Board for Doctorates of Delft University of Technology
and of the Academic Board of UNESCO-IHE Institute for Water Education
for the Degree of DOCTOR

to be defended in public
on Monday 1st of October 2012 at 10:00 hours
in Delft, the Netherlands

by

Carlos Alberto VÉLEZ QUINTERO
born in Cali, Colombia

Master of Science with Distinction in Hydroinformatics
UNESCO-IHE Delft, The Netherlands

This dissertation has been approved by the supervisor:

Prof.dr.ir. A.E. Mynett

Composition of Doctoral Committee:

Chairman	Rector Magnificus, TU Delft
Vice-chairman	Rector UNESCO-IHE
Prof.dr.ir. A.E. Mynett	UNESCO-IHE, TU Delft, supervisor
Em.Prof.dr. R.K. Price	UNESCO-IHE, TU Delft
Dr.ir. A.H. Lobbrecht	UNESCO-IHE, HydroLogic
Prof.dr.ir. F.H.L.R. Clemens	TU Delft
Prof.dr.ir. L.C. Rietveld	TU Delft
Prof.dr. D. Buttler	University of Exeter, United Kingdom
Prof.dr.ir. N.C. van de Giesen	TU Delft (reserve member)

Published by:
CRC Press/Balkema
PO Box 447, 2300 AK Leiden, The Netherlands
e-mail: Pub.NL@taylorandfrancis.com
www.crcpress.com – www.taylorandfrancis.co.uk – www.ba.balkema.nl

ISBN 978-1-138-00002-5 (Taylor & Francis Group)

Preface

This book is the result of 5 years of research and more than 10 years of practical experience in urban wastewater infrastructure and their impacts in the receiving rivers. The motivation of the topic came with the experience gained in projects were urban wastewater systems were analyzed separately. Even though we understood the importance of the interaction between components, the knowledge and tools were not available to put this in practical applications.

This book gives substantial evidence of the importance of the integrated management of urban wastewater systems, if the protection of the receiving waters is to be achieved cost-effectively. This new way of thinking, that considers the dynamic interaction between components, was tested in the design and operation of sewerage networks and wastewater treatment plants (WwTP). This book introduces a method named Model Based Design and Control (MoDeCo) for the optimum design of urban wastewater components. MoDeCo combines the traditional design approach with integrated modelling tools and multi-objective optimization algorithms. This book presents a detailed description of the modelling tools developed. Readers can learn to analyze data and implement an integrated model of the urban wastewater system by following the two case studies presented. This book also presents two alternatives to solve the problem of computing demand in optimization of urban wastewater systems: the use of surrogate modelling tools and the use of Cloud computer infrastructure for parallel computing.

The integrated modelling tools and multi-objective evolutionary algorithms given in this book are excellent tools for researchers and practitioners interested in planning and development of urban wastewater infrastructure. This book aims to help practitioners to optimize the design and operation of urban wastewater systems not only by reducing the cost of the infrastructure but also the risk of flooding and the pollution impacts.

Writing this book taught me that a research is not finished until is written and that is perhaps the most difficult part. To be able to transfer the acquired knowledge into this book was a personal challenge. But making it comprehensive and readable was only possible with the significant effort of my mentors Prof. Roland Price and Assoc Prof. Arnold Lobbrecht. Thank you to their proofreading, editing and advices, this book becomes the instrument to transfer the knowledge and experiences acquired during my research to future researchers and practitioners.

Summer of 2012,
Delft – The Netherlands

i

Acknowledgements

In the first place I would like to acknowledge the support of the institutions that make possible this research. The main support came from UNESCO-IHE the Institute for Water Education who financed this research through the Delft Cluster Project. Acknowledgments are also given to the Water-Board of Rijnland for co-funding this research.

Testing the approach developed in this research with two case studies was possible thank you to the generosity of the institutions that provide the data. The case study in Gouda (The Netherlands) was developed with the data of the wastewater system provided by the Water-Board of Rijnland and the tools and technical support to handling precipitation and forecast data given by HydroLogic BV. The case study in Cali (Colombia) has the generous support of a group of institutions related with the wastewater system. The utility company Empresas Municipales de Cali EICE ESP (EMCALI); two environmental authorities: Departamento Administrativo de Gestión Ambiental de Cali (DAGMA) and Corporación Autónoma Regional del Valle del Cauca (CVC); and two research institutes: Instituto de Hidrología, Meteorología y Estudios Ambientales de Colombia (IDEAM) and Instituto Cinara from Universidad del Valle.

I want to express my special gratitude to my mentors Prof. Roland Price and Assoc. Prof. Arnold Lobbrecht. Professor Price guided me through this research. I had the privilege to have uncountable discussions that helped me to clear the path of my research. He always has great insights and his advices helped me to overcome the technical challenges that the topic impose. I would like to express my enormous gratitude to Assoc. Prof. Lobbrecht. He gave me the opportunity to do this research and was an unconditional support until the end of it. Thank you Roland and Arnold for making this dream come through.

I am indebted to my promoter Prof Arthur Mynett. He was my supervisor during my MSc research and now he helped me in the critical moment when I needed to finalize my PhD research. I would like also to thank other members of the doctoral examination committee for their reviews and comments on this thesis: Prof. F.H.L.R. Clemens, Prof. L.C. Rietveld, Prof. D. Buttler and Prof. N.C. van de Giesen.

During these years I also have the privilege to work with numerous people that contributed in different ways to my research. I would like to acknowledge the help of Prof Dimitri Solomatine. I have to thank him for believe in my ideas and impulse then forward. His support was invaluable to develop the chapter 6 of this book. Thank you also to Dr Schalk Jan van Handel with whom I share the interest for RTC and who inspired me to apply anticipatory control for the wastewater treatment of Gouda presented in chapter 5. I have also the opportunity to work with two MSc students Elena Samitier and Xu Zheng for whom I have my appreciation for their contributions to this book with their researches.

The contact people in the institutions that collaborated with my research were extremely helpful. Here I would like to thank Paul Versteeg and Leo Authier for their support to understand Gouda wastewater system. Thank you to Timmy Knippers for his support with the HydroNET software. Thank you also to all the people that facilitated acquiring data for the case study in Cali within them: Francisco Camacho and Jose Ceron from EMCALI, Oscar Ramires from CVC, Gisela Arizabaleta from DAGMA, Mariana from IDEAM and Alberto Galvis from Cinara.

Optimization of Urban Wastewater Systems using Model Based Design and Control

I want to acknowledge the help and support of my friends and PhD fellows: Arlex Sanchez, Leonardo Alfonso, Gerard Corzo, Wilmer Barreto and Juan Pablo Silva. Their friendship has cross the borders and they have become my brothers and their families my extended family in the Netherlands. With them I have shared the sweet and the sour of this journey. We had endless discussions about projects and inventions, dreams that I hope will be the topic of our next book.

Finally yet very important, I would like to thank my family. I want to thank my partner Ružica who has been supportive and comprehensive on my uncountable nights and weekends in front of the computer. She even expend some time proofreading some parts of this book and commenting it with healthy criticism. Thank you to my parents that during these years have been encouraging me to reach my goals. Thank you to my nice Diana and nephew Andres for being the company of my parents in Colombia. And finally I have to express my special gratitude to my sister Nelcy, who is a wonderful human being and have been the support of my parents in my absence. With out their support I will not be able to achieve my goal.

Carlos Vélez
Delft – The Netherlands

Dedicated to my beloved son
Jovan Vélez Jaćimović

Summary

The pressure on the environment increases as urbanisation continues relentlessly in virtually all cities of the world, and as climate change appears to lead to more extreme rainfall in many urban areas. These developments have an effect on both water quantity and quality in urban drainage systems. Without additional measures, urban wastewater systems (UWwS) will become overloaded more often, generating more frequent flooding events and polluted discharges into receiving waters. A considerable amount of scientific evidence had been collected which leads to the conclusion that the urban wastewater components such as: rainfall-runoff, wastewater from households and industry, storage, pumping, overflows, wastewater treatment and receiving waters, should be treated as one integrated system, rather than separate systems, if the protection of the receiving waters is to be achieved cost-effectively. Even more, there is a need to optimize the design and operation of the sewerage network and wastewater treatment plant (WwTP) considering the dynamic interactions between them and the receiving waters.

This research answers two main questions: first, how to optimize the design and control of UWwSs considering the interaction between the different components? And second, what are the main benefits and drawbacks of this approach? The first question is answered by presenting a methodology called Model-based Design and Control (MoDeCo). The second question is answered through the implementation of the methodology in two case studies: the design of a sewer network in Cali, Colombia, and the functional design of the wastewater treatment of Gouda, The Netherlands.

The Model Based Design and Control (MoDeCo) approach can be described as a combination of the iterative design and model predictive design approaches. Thus, MoDeCo starts with a pre-design that is based on traditional approaches and empirical rules of operation. This is the way in which almost all existing urban wastewater drainage systems have been designed and built. Subsequently, there may be attempts to improve the performance of individual components of the system. The novel approach adopted by MoDeCo is to continue at the design stage with the results of the pre-design in order to build a model of the complete system. This model is used to explore the performance of different designs. Alternative designs are automatically generated and a set of optimum designs is found using multi-objective optimization algorithms. The conceptual framework includes six steps: (i) problem definition, (ii) pre-design of UWwS components, (iii) pre-design of operational strategies, (iv) implementation of the model of the system, (v) optimization using multi-objective evolutionary algorithms and (vi) post-processing of the set of optimal solutions.

To realize the methodology an integrated modelling tool for the UWwS was developed. The integrated model tool consist of three state-of-the-art modelling tools linked together: Storm Water Management Model (SWMM) for the hydrology and transport processes in the urban catchment and sewer network, STOAT dynamic modelling software for the processes in the WwTP and the Water Quality Analysis Simulation Program (WASP) to simulate the processes in the river. Customized algorithms where used to coupling the three software. In addition the integrated model was linked with a multi-objective evolutionary algorithm. The selected algorithm was the Non-dominated Sorting Genetic Algorithm (NSGAII). In general, the modelling tools allow the designer to modify the design parameters and also to generate the data required to calculate indicators related with water quantity, water quality and cost.

The first test of MoDeCo approach was in the design of the sewer network for an area of 70 ha in the expansion zone of Cali (Colombia) that will provide housing for approximately 22000 inhabitants by the year 2030. The urban wastewater infrastructure includes: a combined sewer network, a storage tank, a combined sewer overflow (CSO) network, a pumping station, and an activated sludge treatment plant. Both surface drainage and treated wastewater will discharge to the River Lili. For the case study we pre-design all the components of the UWwS using traditional methods and empirical rules. The sewer network was designed using the rational method and the Colebrook-White formula for the routing of flows. The treatment plant was designed considering the removal of organic mater and nitrogen following the modified Ludzack and Ettinger scheme. The storage volume was estimated and the setting for the operation (control) of the weir for the CSO and the pumping station were defined based on empirical rules, e.g. pump flow during rainfall events equals 2 times the dry weather flow (DWF).

The pre-design of the components and the setting for the control of the ancillary structures where used to schematize the model of the sewer network and the treatment plant. Existing information of the Lili River was used to instantiate the model of the river. The integrated model of the system was used to assess the effects of the design variables in selected performance indicators for flooding on the urban catchment, pollution impacts in the Lili River and costs. The performance of the pre-designed system shows that the system is capable of protecting households from flooding for the design rainfall event with a return period of 20 years. However, performance in terms of pollution impacts was poor, showing an oxygen deficit in the river in the order of 6.2 mg/l, which implies that the minimum dissolved oxygen (DO= 1.6 mg/l) was well below the standard defined for the river (4 O_2 mg/l).

The optimization of the sewer network was posed as a multi-variable and multi-objective problem in which the aim was to find the combination of pipe diameters, storage volume and pumping set points that minimize the flooding volume, the pollution in the river measured as deficit of DO and cost of the system designed. With the MoDeCo approach it was possible to analyze around 50,000 design alternatives and to come up with a handful of Pareto set optimum solutions. On average, it was possible to optimize the sewer network design and reduce the cost on average up to 15% when compare with the pre-designed system, maintaining the same level of protection against flooding. The best design alternative seems to require an increased storage volume of 3 times the pre-designed volume and to set the pumping station to a maximum capacity of 5.5 times the DWF. This will increase the overall cost of the system by 35% when compared with the pre-design, but reduce significantly the water quality impact in the receiving system. The minimum DO with the best alternative design is 4.1 mg/l which is above the standard required for the River Lili.

Perhaps, the greatest advantage of the MoDeCo approach for sewer design is that the alternative solutions correspond to an integrated analysis of the system in which the synergy between the three main components of the system: sewer network and ancillary structures, wastewater treatment plant and the river have been included. The obtained solutions are optimal not only for protecting the community from flooding events but also for protecting the environment that receives the discharges from the city.

The second case study is used to test the MoDeCo approach for an existing UWwS. It also focuses on optimizing the control of the system. The case study is developed for the UWwS of the city of Gouda in the Netherlands, a system that serves a population of approximately

71,000 inhabitants. The city has 12 drainage areas and the sewer system is evolving from a combined to a separate drainage scheme. The WwTP is designed for biological removal of organic matter, nitrogen and phosphorous components. The combined sewer overflows (CSOs) are discharged onto open surface canals that serve as city drainage. The final discharge of the surface canals is in the Hollandse IJssel. This river also receives the treated effluent of the WwTP. The main concern of the Rijnland Water Board (WwTP operator) is to comply with stricter effluent quality standards for total nitrogen (Ntot-N ≤ 5 mg/l) and total phosphorous (Ptot-P ≤ 1 mg/l). The aim of the case study was to demonstrate the benefits of applying MoDeCo approach to design a better control/operation strategy for Gouda WwTP.

To apply MoDeCo approach, an integrated model of the sewer and the treatment plant was developed. Dry weather flow and rainfall run-off processes were modelled using SWMM. The wastewater composition in the outflow of the sewer was modelled based on curve fitting and M5 model trees. The water quality components were correlated with intensity of precipitation, flows in the sewer, temperature and season. Models for temperature (T), chemical and biochemical oxygen demand (COD and BOD), suspended solids (TSS), total Kjeldahl nitrogen (TKN) and total phosphorous (Ptot-P) were developed. The biological processes of the wastewater treatment plant were modelled using the activated sludge model (ASM2d) implemented in STOAT. Settlers, thickeners and sludge handling were also included in STOAT to have a complete model of the WwTP. Since the main objective was the functional design of the system, proportional and integral (PI) controllers were used to simulate the current operation of aerators and pumps for recycle flows.

The functional design can be summarized as the selection of set points for the operational variables: internal recycle (Qir), air flow rate (Qair) and the dose of readily biodegradable organic mater (VFA). The problem was posed as a multi-objective optimization in which the aim was to find the combination of set points for Qir, Qair and VFA that minimize the effluent concentration of nutrients (Ntot-N and Ptot-P) and the operational cost. The optimum design was defined for different conditions that may disturb the system operation. Therefore the analysis was done considering variations of dry and wet weather flows and wastewater compositions and different temperature conditions (winter and summer). As a base scenario the performance indicators were estimated using the current set points of the Gouda WwTP.

In general, the results of the case study show that is possible to improve the performance of Gouda UWwS by optimizing the functional design. With the operational variables, it was possible to reduce the effluent concentration of total nitrogen while keeping the concentration of total phosphorous below the standards. For instance, for DWF in winter, the concentrations of Ntot-N and the Ptot-P were reduced by 51% and 53% respectively when the performance of the system with optimized set points is compared with the base scenario defined above. The performance of the system with respect to operational cost decreased, but that was expected because the objectives are contradictory (i.e. decreasing the effluent concentrations implies an increase in cost). However, the estimated costs do not include the possible cost savings associated to the reduction of pollution impacts in the river. Perhaps the main benefit of the approach is the generation of new knowledge about the behaviour of the system. For instance, the ratio of Qir/Qin tends to indicate that the optimum set points are more dependent on the flow conditions than on the temperature conditions. From the practical point of view this implies that a set of ratios for different influent flows may help the operators to set up their internal recycling.

One of the main limitations for the use of MoDeCo is the long computing time required to find optimum solutions; a step forward to solve this problem was done by developing and testing two alternatives to reduce the computing demand: parallel computing using Virtual Clusters in the Cloud and a new surrogate modelling method here named the Multi-objective Optimization by PRogressive Improvement of Surrogate Model (MOPRISM). Overall, the experiments presented in this thesis show that there are significant reductions in computing time using surrogate optimization or parallel computing. The advances in surrogate optimization and parallel computing are promising, and the benefits are so important that it is possible to anticipate that in the near future all multi-objective optimizations will include one of these two approaches, or may even be a combination of them. The general conclusion is that designs found with the MoDeCo approach have better performance than designs with traditional methods.

The main benefits of the MoDeCo approach can be summarized as follow:
- the design includes the interaction between all components of the UWwS; this allows designers to expand the scope of UWwSs design to include pollution impacts in the receiving system;
- the use of state-of-the-art modelling tools allows the design to be based on dynamic conditions contrary to traditional approaches that are usually based on constant design flow and fixed compositions of water quality;
- the use of global optimization algorithms increases the chance of finding acceptable optimum solutions;
- the design is driven by the minimization of cost while maintaining the performance of the system for other objectives;
- the approach benefits from the analysis of a great number of alternatives;
- the designer and decision makers are better informed about the solutions and their consequences;
- the MoDeCo approach is in line with new regulations such as the European water framework directive (WFD) that enforces a holistic view of the urban wastewater management and the reduction of pollution impacts on the receiving waters.

The limitations of the approach can be summarized as follow:
- there is often a lack of adequate information to build integrated models of UWwSs. The optimum design depends on the accuracy of the model's predictions, and the uncertainty in the model may threaten the validity of the optimization process. However, the lack of information and the uncertainty in the data are not exclusive limitations of MoDeCo but of any approach that employs a holistic view of the system;
- the combination of integrated models with multi-objective optimization algorithms makes the approach computationally demanding. This may limit the use of the method in practical applications. However, this research shows that parallel computing and surrogate modelling are excellent alternatives to overcome this problem;
- the integration of modelling tools and optimization algorithms may require additional skills from the traditional designers, which may limit the practical application of the approach. However, research like the one presented here, may help to bridge the gap between theory and practice.

In this research, the scope of the design of UWwSs has been expanded to include the dynamic interactions between individual components. Further research should consider long term objectives, for example indicators of sustainability, resilience or robustness and use them to evaluate the optimum solutions found in a post-processing step of the design.

Table of Content

1 Introduction and Scope

1.1 General Introduction to the Research Area

Water managers face new challenges in meeting the requirements of various water uses in urban environments. The pressure on the environment increases as urbanisation continues relentlessly in virtually all cities of the world, and as climate change appears to lead to more extreme rainfall in many urban areas. These developments have an effect on both water quantity and quality of urban water systems. Without additional measures, urban wastewater systems (UWwS) will become overloaded more often as a result of excessive rainfall, with more frequent flooding and polluted discharges from sewers, meaning greater amounts of sewage on the streets. At the same time the receiving surface-water systems will be subject to larger rainfall-runoff from urban areas and from sewerage overflows, leading to pollution of our living and natural environments.

The pressure is bigger when the UWwS has to comply with new regulations such as the EU Water Framework Directive (WFD) (CEC 2000). WFD emphasizes the conservation of good chemical and ecological quality of receiving systems and in that way force to address the urban wastewater management in a more integrated approach. In fact, urban wastewater managers are being forced *to optimize the design and operation of UWwSs* in order to deal with the increased regulatory pressure and new criteria for performance.

1.1.1 Pressure over wastewater systems

The first challenging factor for UWwS is the rapid growth of the urban population. The United Nations (UN) World Population Prospects shows that virtually all the population growth expected between 2000 and 2025 will be concentrated in urban areas. Figure 1.1 presents some of the characteristics of the World Urbanization tendencies based in the world population online data base (UN 2007). Even though in developed countries the already highly urbanized areas are not expected to grow substantially, the less developed countries are expected to have the size of their cities increase significantly between 2000 and 2025. Around 80% of the world's urban population in 2025 will live in developing countries.

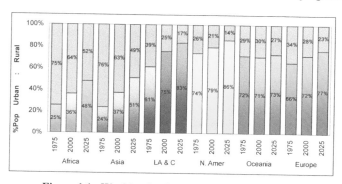

Figure 1.1. World urbanization prospect (UN 2007).

These projected urban population growth figures suggest that urban services will face great challenges over the coming decades to meet the fast-growing needs. Figure 1.2 highlights the challenges faced by the sector in reducing the coverage gap. By the year 2000, 47% of the population in Africa, Asia and Latin America were lacking sanitation services. To reach the Millennium Development Goal (MDG) target of universal coverage by the year 2025, almost 4 billion people will need to be served with sanitation (WHO, *et al.* 2000).

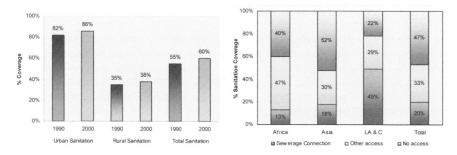

Figure 1.2 a. Global sanitation coverage and b. Sanitation coverage by category of service

Source: (WHO, et al. 2000)

The second factor considered is the climate variability. The effects of large urban areas on local microclimate occur because of changes in the energy regime, air pollution, air recirculation and release of greenhouse gases. These factors change the amount of precipitation and evaporation. Geiger *et al* (1987) in Marsalek *et al*, (2001) state that in large industrialized cities precipitation is 5 – 10% higher than in the surrounding areas and for individual storms, the increase in precipitation can be as high as 30%. This local climate change in addition to the global climate change implies that urban wastewater systems will have to deal with more extreme rainfall events that also impact on the amount of pollution that affect receiving waters in urban areas through combined sewer overflows particularly through the first flush of pollution from drainage pipes and channels.

One of the consequences of wastewater discharge into water receiving systems is the deterioration of water quality. Although there are some clear impacts of urban wastewater systems, it is not easy to assess the impact on water quality in receiving waters at the global scale. This is due to the lack of monitoring capacity and the inherent complexity of both natural and anthropogenic pollutants. However, some of the patterns and trends presented in the scientific literature support the impairment of the receiving waters due to urbanization. One of the global water quality assessments carried out and updated by Meybeck (2003) analyses eleven variables, ranking their effect on the provision of freshwater services. Figure 1.3 shows the general tendencies for specific pollutants, but a wide range is noted, with minima in all cases ranked zero and maxima often several times more severe than the mean condition. The results show that pathogens and organic matter pollution (from sewage outfalls, mainly) are the two most pressing global issues, reflecting the widespread lack of wastewater treatment (Hassan, *et al.* 2005).

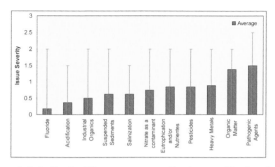

Figure 1.3. Ranking of globally significant water quality issues affecting freshwater resources.

Note: Severity values are as follow: 0: No problem or irrelevant; 1: Some pollution, water can be used if appropriate measures are taken; 2: Major pollution with impacts on human health and/or economic use or aquatic biota; 3: Severe pollution – impacts are very high, losses involve human health and/or economy and/or biological integrity (Bars represent the average severity based on expert opinion and updated by Meybeck, 2003).

Although these updated results correspond to the state of water quality in the 1980–90s (Meybeck 2003), since the 1990s the situation in most developing countries and countries in transition has most likely become worse in terms of overall water quality. In Eastern Europe, Central and Southern populated Americas, China, India, and populated Africa, it is probably worse for metals, pathogens, acidification, and organic matter, while for the same determinants Western Europe, Japan, Australia, New Zealand, and North America have shown slight improvements. Nitrate is still generally increasing everywhere, as it has since the 1950s. In the former Soviet Union there has been a slight improvement in water quality due to the economic decline and associated decrease in industrial activities (Hassan et al., 2005). Data collected and analysed by the Global Environmental Monitoring System on Water (GEMS/Water) over the last two decades for biological oxygen demand (BOD), nitrates and phosphates support the previous analysis presented by Meybeck (2003) and Hassan et al, (2005).

Another consequence of the failure of UWwS to control the pollution is the inability to fulfil rules and regulations. One example is presented in Figure 1.4. Although in many of the European countries the UWwSs are fully developed still the situation of the surface waters is worse than expected according to the first stage report on the implementation of the WFD. Figure 1.4 shows the evaluation of the first stage of implementation of WFD. Most of the surface water bodies EU communities are at risk of failing the good quality and ecology objectives; surprisingly The Netherlands is one of the countries at most risk of failure (CEC 2007). As a consequence of failure to meet the requirements of the regulations, there is a loss of credibility of the urban wastewater management institutions in the eyes of the communities and stakeholders, and an increased requirement to upgrade the UWwS involving costly investments in infrastructure.

Thus, new urban wastewater systems must be developed and existing systems optimized such that they can cope with the growing service demand, highly variable operational conditions and stricter regulations. Therefore there is no doubt that even though there has been lot of research and effort in optimizing urban wastewater systems, the topic continues to be relevant and important in order to achieve a more sustainable development of urban water resources. Before, introducing previous research in this area, some working definitions of the components of UWwSs are given in what follows.

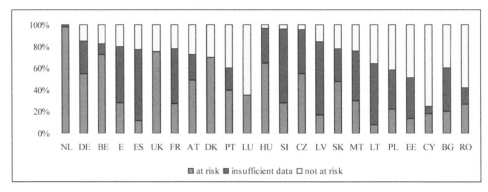

Figure 1.4 Percentage of surface water bodies at risk of failing WFD objectives per European Union member state

Source: (CEC 2007)

1.2   Urban Wastewater System Definitions

Urban wastewater is defined here as domestic wastewater or the mixture of domestic wastewater with industrial wastewater and / or storm-water run-off. This definition is based on the Council of the European Communities Directive concerning urban wastewater treatment (CEC 1991). The expressions, "wastewater" and "sewage" are rather confusing because they may or may not include run-off. The first one is used in USA; the second one is more common in UK, but hereafter we use them as synonyms. Distinction is made between run-off sewage and sanitary sewage; the latter stands for domestic and industrial wastewater.

An Urban Wastewater System (UWwS) is composed by the Sewer network, the Wastewater Treatment Plant (WwTP) and the Water Receiving System (e.g. rivers or lakes). The UWwS has links with other urban water components like groundwater, rural streams, drinking water production and supply, and agricultural runoff. However, the focus of this research is on the three subsystems included in the definition above and shown in Figure 1.5.

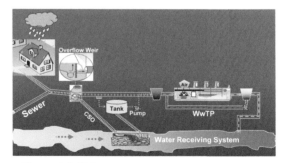

Figure 1.5 Urban wastewater system components

1.2.1 The sewer network

Sewer networks (or sewerage systems) consisting of open channels and pipes are used to collect and convey both run-off and sanitary wastewater out of urban areas. These sewer networks transport the water either directly to the receiving water or to a wastewater treatment plant. Generally speaking two different types of sewer systems can be found:

- Separate sewer systems use two separate conduits to convey the sanitary wastewater and stormwater:
  - Sanitary sewers, which drain wastewater from households, industries and public buildings. They drain the sanitary sewage to a wastewater treatment plant or directly to the receiving system in case of a sanitary sewer overflow (SSO).
  - Storm sewers that collect precipitation that falls on the urban catchment area and does not infiltrate into the ground or evaporate. This stormwater runs off the surface from which organic and inorganic material is lifted up and transported along with the water. The drains convey the run-off to the closer open surface system (consisting largely of urban streams) or to storage tanks before is discharged to the receiving water system.
- Combined sewer systems are characterized by the use of only one conduit where sanitary sewage and run-off sewage are mixed and transported together. If the flow in the sewer system becomes greater than the hydraulic capacity of the pipes or the WwTP, the water leaves the system via emergency exits, or combined sewer overflows (CSOs)

Other classifications may consider for instance gravity sewers, and pressurized sewers depending on the driving force of the water. For this research we use as an experimental subject a combined sewer, but techniques developed are also valid for other types of sewers.

1.2.2 The wastewater treatment plant

The wastewater treatment is the process of removing contaminants from the wastewater in order to produce an effluent that does not adversely impact the quality of the receiving system and its uses. The treatment includes physical processes (e.g. sedimentation or filtration), chemical processes (e.g. precipitation or flocculation) and/or biological processes (e.g. aerobic or anaerobic degradation of organic mater by bacteria) (Tchobanoglous, *et al.* 2003). Three levels of treatment are distinguished: primary, secondary and tertiary treatment. The level of treatment depends in general on the size of the urban population (generated load), the type of receiving water body (sea, river, lake, estuary), the water quality requirements of the receiving body (sensitivity of the area, downstream uses, etc), and the requirements of the water legislation.

- Primary treatment means treatment of urban wastewater by physical and/or chemical processes involving the settlement of suspended solids in which the organic mater (measured as biochemical oxygen demand BOD_5) is reduced by at least 20% and the total suspended solids (TSS) are reduced by at least 50% (EEC 2007). Typical process units included at this level of treatment are: screening, grit chamber and primary sedimentation tank (coagulation and flocculation are used to enhance settling ability of the suspended solids).

- Secondary treatment means a treatment of wastewater that include biological processes and secondary settlement in which the organic mater measured as BOD_5 is reduced between 70 to 90%, the chemical oxygen demand (COD) is reduced at least 75% and TSS are reduced between 60 to 90% in relation with the influent load (EEC 2007). The core process unit in secondary treatment is the activated sludge that uses aeration and agitation to facilitate the conditions for growth of heterotrophic microorganisms that degrade the organic mater. Other process units used for secondary treatment are: trickling filters, anaerobic reactors and stabilization ponds.

- Tertiary treatment provides a final stage to raise the effluent quality before it is discharged to the receiving environment. The removal of nitrogen (N) and phosphorous (P) are within the process commonly added to the secondary treatment. The removal of nutrients is achieved using chemical or biological process. Common removals efficiencies are above 80%. Other tertiary treatments like sand filtration and disinfection are always at the end of the process and are also called effluent polishing.

For this research we use as an experimental subject the activated sludge treatment processes with nutrient removal. Figure 1.6 shows a scheme of an activated sludge plant with biologic nitrogen removal. This is one of the systems most used in urban areas.

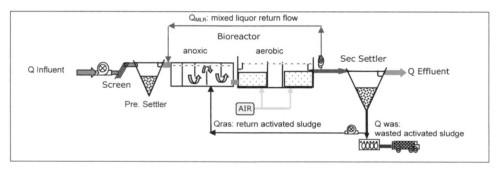

Figure 1.6 Wastewater treatment plant scheme.

1.2.3 The receiving water system

The urban wastewater is mainly discharged into natural surface water bodies nearby the urban area. In general, four types of receiving water bodies can be distinguished: rivers, lakes, estuaries and costal areas. Within them, natural drainages (rivers) are more frequently used as discharge point for storm sewer outfalls, CSOs and effluents from WwTPs. In addition to serving as natural drainage for urban wastewater, the receiving water bodies also have other functions such as: transport, recreation, fishing, drinking water production, irrigation and habitat for aquatic ecosystem. All these functions can only be maintained by the combination of two factors: the quantity and the quality of the water in the water body. Only if the right quality of water is present in sufficient quantity, can all the functions be supported.

In order to preserve the functions of the receiving system, the sewer network and the WwTP have to be designed and operated in a way that limits the adverse impacts that the discharges may cause on the receiving waters. Wastewater discharges can generate morphological changes; deteriorate the water quality and impair the aquatic ecosystem. The impact of the wastewater discharge not only depends on the wastewater characteristics but also on the type of receiving system and its hydrological, morphological, water quality and ecological characteristics. Some of the receiving water body characteristics that may influence the decision of allow an urban wastewater discharge are (House, *et al.* 1993):

- The extent, sensitivity and importance of the local aquatic life.
- The identification of key pollutants (e.g. biodegradable organic matter for rivers and nutrients in lakes, estuaries and coastal areas).
- The transport and dilution of pollutants (e.g. flowing, semi-stagnant and stagnant waters)
- The auto-depuration and assimilation of toxic pollutants (e.g. oxidation of organic mater).

1.3   Design of Urban Wastewater Systems

Although there are different options of sanitation in urban areas, the focus of this research is on those urban systems that use or are suitable to use sewers for the collection and transport of wastewater. Thus, the design from now on refers to these components: sewers, storage tanks, wastewater treatment plants and ancillary structures (overflows, pumps, etc) taking into account the impact on the receiving waters. The sewer design may be divided into two phases: (i) Selection of network layout and (ii) hydraulic design of the sewer pipes in the selected layout (determination of discharge rates, pipe sizes, slopes, and invert elevations). The WwTP design can be divided into two phases (i) Process design (determination of volumes of reactors for liquid phase and solid phase) and (ii) hydraulic design (flow splitters, pipes, pumps, etc). Both components require a functional design that is in this research understood as the definition of the operational set point of the components. According to Harremoës and Rauch (1999), there are two extreme approaches for the design of the components: *the empirical iterative approach* and *the prediction-design approach*.

1.3.1   Empirical iterative approach – static design

In the empirical iterative approach, structures for pollution abatement are built on simplified assumptions and their performance is subsequently evaluated through monitoring. When the monitoring system proves that the performance is inadequate, then an improved plan of action is implemented. This approach is significantly different to the second approach by advocating a purely inductive interpretation of information from experience gained by operating the systems in question (and other systems), from which a pattern can be identified and responded to in an empirical, iterative approach to design and operation (Harremoës and Rauch 1999).

The experience gained and patterns identified became design rules that are typically used for setting up urban water facilities and could be considered as the oldest and simplest models. Those rules summarize experience and conceptual thinking, and they have proved their usefulness. Depending on the school to which designers belong, different static models are used and modifications applied as knowledge of the processes increases. The rational method is one example of this approach explained by Vanrolleghem and Schilling (2004):

> "The rational method, developed 150 years ago, is a way to give an estimate of the safe side of the maximum runoff Q_m from an urban catchment given a constant rainfall intensity i: $Q_m = ciA$. Together with a time-offset model for routing the flow (constant flow velocities in all conduits), this is a simplified formulation of the rainfall-runoff process".

Thus, with some safety factors, these simplified models are often applied for design purposes. However, if these design models are applied to cases outside the range of their validity errors can be made. In the case of the rational method / time offset model for urban drainage, this works relatively well in systems with moderately steep, dendritic (tree-like) urban drainage systems for intensive rainfall. If this kind of model is applied to a flat system with a looped sewer network in a less urbanised catchment during winter situations the results will deviate significantly from what works in reality. It would not be possible to validate this model for such a system (Vanrolleghem and Schilling 2004).

A factor that limits this static approach is climate change and its direct impact on precipitation patterns. One example is presented by Arnbjerg-Nielsen (2006), who found a significant trend in the 10 minute maximum intensity in the eastern part of Denmark, indicating that existing intensity-duration-frequency (IDF) curves are today no longer valid for design practice as precipitation pattern changes have already occurred in the region. Another example is given by Grum et al (2006), who conclude that with the changed rainfall patterns it will be necessary to rethink the current design criteria. These authors suggest that new designs should not be related only to the frequency of a given occurrence such as flooding but also to criteria that lead to the appropriate handling of all extreme events, including those that cannot be contained within the traditional drainage system (Grum, et al. 2006).

Another factor that limits the static approach to design is the need for optimising the efficiency of the UWwS with regard to the ecological consequences in natural water bodies and with regard to the investment and operation costs as WFD request. With this new water-quality based approach, the design is far less predetermined and the options to meet the goals become much more widespread (Benedetti, et al. 2004b).

1.3.2 Prediction design approach – dynamic design

In the prediction-design approach, models play an essential role in the prediction of performance and the evaluation of competing design alternatives. This approach is dominated by a deductive interpretation of the problem. It is based on the idea that if the problem is reduced to its basic components and tied together in a system of physical, chemical and biological laws of natural science, the future can be predicted with sufficient accuracy to warrant a safe design and operation. The idea is to identify a set of laws to be used in a model structure based on prior knowledge, and to calibrate the unknown parameters against data from reality (Harremoës and Rauch 1999).

In principle, the prediction approach has more universal applicability than the empirical approach because it looks for the cause – effect relationships through investigations and monitoring. However, in the end the predictive approach cannot avoid significant elements of pragmatism, because investigations and monitoring provide the empirical basis for the structure of the reasoning and parameters of the models (Harremoës 2002).

Thus, there are some drawbacks in the prediction approach. Harremoës and Rauch (1999) for example mention the failure to model the cause – effect relationships. In general, there exist limitations in the modelling when it comes to water quality or ecology issues, for example, the model pollutant transport during rainfall-runoff or the relationships between intermittent loads on rivers due to rainfall-runoff and the resulting effect on the ecosystem (Harremoës 2002).

Harremoës and Rauch (1999) suggest that further development should be a combination of elements from both approaches. Thus, a static design and then a dynamic assessment is perhaps one of the most common ways to include models in the design. First, the component (Sewer network or WwTP) is designed for steady state conditions and then the model is built to assess the functioning of the system under different scenarios. In fact, models are not generally developed to design the systems, only to assess their performance under steady state or dynamic conditions.

1.4   Integrated Urban Wastewater System

A conventional practice has been to design and operate the sewer and the treatment plant in isolation. For example, the study of various design options for the sewer system often ends at the overflow structures and the treatment plant inlet, whereas the role and function of the treatment plant and the receiving water body should also be taken into account (Butler and Schütze 2005). The transfer across the interfaces of each subcomponent is characterized by static rules. For example, the flow to the WwTP under wet weather conditions is limited to a value the order of twice the peak discharge for dry weather conditions, or the number of CSO discharges per year is restricted to a certain frequency (Rauch, et al. 2005).

The concept of an Integrated UWwS is not new; since the late 1970s the watershed-wide planning philosophy has gained attention. In 1992, the United Nations Conference on Environment and Development established the basis of the Agenda 21 and in it defined the principles and guidelines for sustainable urban water management in which the concept of the Integrated UWwS management was strengthened. The first conference on Interactions between sewers, WwTP and receiving waters in urban areas – the InterUrba workshop - held in 1992, was a determining step forward to promote the integrated approach (Harremoës, 2002). More and more scientific arguments have been put forward to state that it is necessary to consider the urban wastewater components as one integrated system rather than separate systems if the protection of the receiving water is to be achieved cost-effectively (Harremoës and Rauch 1996, Lobbrecht 1997, Schütze 1998, Clifforde, et al. 1999, Meirlaen 2002, Langeveld 2004, Vanrolleghem, et al. 2005).

The inclusion of the receiving water characteristics in the design and operation of UWwSs bring as a potential benefit less prescribed sewer and treatment systems and open up more options to meet the goals of the system. Thus, the interactions between the sewer system, WwTP and receiving water body, as well as between different measures to optimize the system components, may result in synergy effects that benefit the overall performance of the UWwS (Benedetti, et al. 2004b).

Some national regulations already include a more integrated approach for the control of pollution, such as the Urban Pollution Management Manual (UPM - 1994) in UK. Although it is a planning guide more oriented to wet weather discharge design, its approach considers the impact on the water quality in the receiving system to assess control options. The procedure involves four main phases: A. initial planning, B. assembling data and tools, C. developing solutions and consenting and D. detailed design (FWR 1994). The potential interaction with the WwTP is poorly considered because this guide focuses mainly on design CSO structures and retention tanks. The pollution impact on the receiving water is considered via simplified models and statistical evaluation of extreme values, mainly focusing on oxygen depletion and on ammonia concentrations in the receiving system. In addition, the allowed impact to the river is defined via duration/frequency curves of certain concentrations (Rauch, et al. 2005). UPM mainly recommends detailed models when the urban catchment population is bigger than 20000 and/or when the interaction with the other components is significant.

In the European WFD the water quality-based approach to manage the pollution in urban systems demands the evaluation of the cause–effect relationships between loads from the wastewater system and the effects on the receiving water. In addition, this water quality-oriented approach offers greater degrees of freedom for improving the wastewater system's performance, because the choice of measures is not constrained by prescribed guidelines. Thus, the potential synergy originating from the interactions between the subsystems may be beneficially used to reduce the pollution impact. In conclusion, as mentioned by Harremoes and Rauch (1999), **there is a need for the design and operation of sewer system, treatment plant and water receiving systems to be done in an integrated way**. There is also a need for design and operation to be based on a more realistic set of water quality criteria to be met by the performance of the system in total. This implies the use of integrated modelling tools to assess competing design alternatives in a dynamic manner.

1.5   Problem Statement and Scope of the Research

More and more scientific arguments are put forward to state that there is a need to consider the UWwS as one *integrated system*, rather than as the ad-hoc combination of separate systems, if the protection of the receiving waters is to be achieved cost-effectively. Therefore, the scope of this research is to contribute to the reduction of urban pollution affecting receiving water systems through the optimization of the design and control of the integrated UWwSs. This research addresses two main questions: first, how to optimize the design and control of the UWwSs considering the interaction between components? And second, what are the main benefits and drawbacks of this approach?

1.6   Outline of the Chapters

The general structure of the thesis is presented in the Figure 1.7. The document contains 7 chapters: the present chapter is the introduction, a chapter with the literature review, a chapter with the methodology proposed, two case studies, a chapter with advanced research on computing time reduction for the optimization process and the final chapter with conclusions and recommendations. A description of the structure of each chapter and the goals is presented in what follows.

- Chapter 1 Introduction and scope: describes the relevance of the research topic, introduces and reviews previous research in the area and establishes the research niche by indicating the gaps in previous research, and outlines the present research.
- Chapter 2 State of the art in the optimum design of urban wastewater systems: this chapter presents and reviews details of previous research in optimization of UWwSs. The goal is to present the conceptual bases required to develop the framework for the model-based design and control of UWwSs.
- Chapter 3 Framework for the optimum design and control of urban wastewater system: this chapter proposes the general methodology for the Model Based Design and Control (MoDeCo) of UWwS. The goal is to describe the approach, and the information and tools required to implement it.

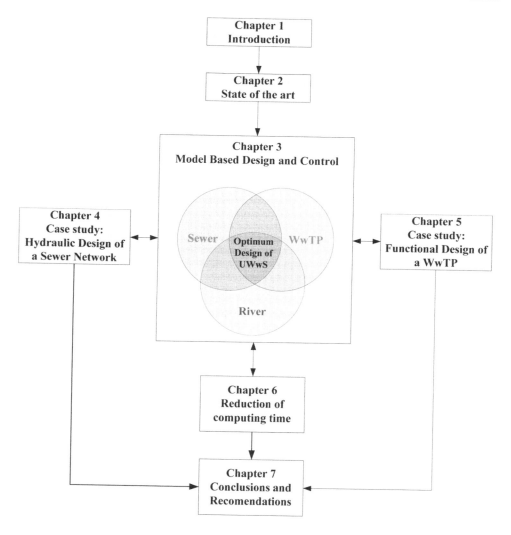

Figure 1.7 General structure of the thesis

- Chapter 4 Design of an urban wastewater system for Cali: in this chapter the approach proposed in chapter 3 is implemented in a case study for the design of a sewer network. The chapter can be divided in three sections, first the case study and the information available is described, second a description of the modelling tools developed is presented and third the optimization results are presented and discussed. The goal is to use the case study to find out the benefits and drawbacks of the approach proposed.

- Chapter 5 Functional design of Gouda wastewater treatment plant: in this chapter the approach is implemented to optimize the operation of a WwTP. The chapter is also divided in three sections, namely descriptions of the case study, modelling tools and results and discussion. An exploratory study of anticipatory control is included at the end of this chapter.

- Chapter 6 Use of cloud computing and parallel computing in optimization processes: This chapter addresses solutions for one of the main problems of the approach, namely the high computing demand of an optimization process. It explores two alternatives to reduce the computing time in an optimization process of UWwSs and can be seen as a complement to the framework proposed.
- Chapter 7 Conclusions and recommendations: this final chapter summarizes and discusses the main findings. It includes recommendations for practitioners interested in practical applications and for researchers interested in exploring further the whole aspect of the optimum design of UWwSs.

2 State of the Art in Optimal Design and Control of Urban Wastewater Systems

2.1 Optimal Design of Urban Wastewater Systems

Designing sewer networks can be a time-consuming task, particularly when the design is largely based on trial and error where suitable pipe diameters and slopes combinations for all pipelines between manholes must be identified. Since there is a large range of possible slopes, diameters and roughness coefficients of pipes, only a small number of combinations of these parameters are usually analyzed during a design process. The design of a wastewater treatment plant also faces the same problem. Usually a small number of combinations of loading rates and mean values of water quality in the influent are used to estimate the volumes of the reactors. In the case of activated sludge process, the possible combinations of internal recycle sludge, chemical dosage and aeration rate are enormous. Thus, the design, even using modelling software to assess the alternatives, ends in the evaluation of a few scenarios that will depend on the expertise and skills of the designer and normally result in sub-optimal designs.

Identifying a minimum cost design is an important issue when constructing sewer networks or wastewater treatment plants. This is perhaps the main driving force of the optimization of for instance storm drainage system (Guo, *et al.* 2008). An optimum design must fulfil the objectives of the system (e.g. prevent flooding and limit pollution impacts) in a cost–effective manner. Therefore, an optimization process requires not only the use of modelling tools but also optimization algorithms that can assist the designer in the search of the optimal solution of the problem.

2.1.1 Optimal design of sewer networks

In contrast to traditional approaches (i.e. empirical iterative approach) in an optimal design of a sewer network the design is treated as an optimization problem. Thus, the designer aims to minimize the construction and/or operational cost whilst maintaining the degree of performance required for the system under certain constraints. In the past four decades, the optimal hydraulic design of a branched sewer system has been the topic of many researchers. A comprehensive literature review on optimal design of storm sewer networks can be found in Guo *et al.* (2008). These authors describe five main advantages of optimal design when compared with traditional approaches:

- The cheapest design solutions may be obtained while providing more reliable serviceability
- It enables sewer engineers to assess the performance of a great number of design alternatives
- Sophisticated models of the system can be used to assess the performance under dynamic conditions
- Local economic conditions can be incorporated in the hydraulic design of the system
- It can ease the design process by automatic computer-based design

Various optimization techniques have been used including Linear Programming (LP), Successive Linear Programming (SLP), Non-Linear Programming (NLP) and Dynamic Programming (DP). However, according to Guo, *et al.* (2008) they have limited success due to strict requirements, for instance: i) for LP the objective functions must be linear or differentiable which is hardly the case in a sewer design (Dajani and Hasit 1974), ii) for NLP, variables are treated as continuous, so there are limitations when considering the sizes of pipes that are discrete variables (Price 1978), iii) DP treat the optimization as a sequential problem, therefore pipe sections are optimized without the possibility to consider the upstream effects of backwater and limiting the assessment of flooding events (Mays and Yen 1975, Gupta, *et al.* 1983). Perhaps a more successful approach is presented by Lobbrecht (1997) that demonstrate that SLP is faster than linear programming and does not has the limitation of the linearity.

With the development of random search methods, a lot of research has been carried out. Within the randomized search methods, the genetic algorithm (GA) appears to be the most used and successful optimization method for the optimal design of sewer networks (Nicklow, *et al.* 2010). Numerous examples of GA application to optimize the hydraulic design of simplified sewer networks can be found in the literature (Cembrowicz 1994, Parker, *et al.* 2000, Afshar, *et al.* 2006, Barreto, *et al.* 2006, Farmani, *et al.* 2006). Other advanced search methods have being implemented like Max-Min Ant System (Afshar 2006) and Particle Swarm optimization (PSO) (Afshar 2008); however, these methods generally encompass similar features to those of GAs and the optimization efficiency and effectiveness may be greatly reduced when handling large networks.

Resent research has looked at the design of the sewer network considering both the optimal layout and the optimal hydraulic performance. For this purpose, hybrid approaches that combine the best features of more than one optimization method have become increasingly used. For instance, Diogo *et al* (2000) combine the DP technique with simulated annealing and GA in a discrete combinatorial optimization problem. Weng, *et al.* (2004) combined the Mixed Integer Programming (MIP) with GA to solve the optimal layout and optimal hydraulic design problem. Other authors have used the hybrid approach to speed up the optimization process by combining GA with, for instance, Tabu search (Liang, *et al.* 2004) and Cellular Automata (CA) (Guo, *et al.* 2007).

Even though there has been a significant contribution from the researchers is this area, putting into practice the optimum design of a sewer network appears to be far from realization. Some of the reasons given by Guo, *et al.* (2008) are:
- The problems concerned appeared to be too trivial, such as designing a small network with a limited number of decision variables and a single objective. Even more, the objectives of sewer design has been expanded beyond those of cost and flood protection, by involving environment, ecology, energy, sustainability, maintenance, control and management interests. However, optimal design considering for instance pollution impacts is currently at its very early stage and certainly requires considerable further research.
- The optimization of the design of a sewer network is fundamentally a multi-objective problem (i.e. minimization of cost, flooding and/or pollution impacts). However, it is common practice to treat the problem as single objective by using weighting factors or by transforming objectives into constraints.
- Although sophisticated model of the sewer network exist, highly simplified hydrological and hydraulic models are generally used in optimization practices.

- One significant limitation when applying optimization algorithms is that these techniques entail an excessive computation time. A trade-off is always necessary between affordable computational cost and acceptable solution optimality, especially for big or complex design problems. Moreover, in most cases, only near-optimal solutions can be guaranteed instead of obtaining true optimal or globally optimal designs.
- Costing models employed in sewer optimization were often highly simplified, generalized or unrealistic; hence the costs produced by these models can notably differ from the actual tender costs. As a result, the solution obtained by design optimization may deliver unreliable cost information, and even not be optimal in reality.
- High uncertainties generally exist in system loads, hydrological and hydraulic processes, and modelling. The virtues of precisely cost-effective designs, obtained by optimization under certain design criteria, are highly compromised by these uncertainties, especially for designs without consideration of reliability and risks.

2.1.2 Optimal design of wastewater treatment plant

The optimal design of a WwTP aims to minimize the construction cost and/or the operational cost whilst maintain predefined performance criteria; for instance, the efficiency of removal of organic mater and nutrients. In general, an objective function is formulated that serves as the criterion in determining an admissible set of processes variables and operating conditions. Both the objective function and the constraints (the equations and inequalities) must be expressed in terms of variables representing measurable or controllable variables in the process (Himmelblau 1976).

Similarly to the optimum design of sewer networks, the research on optimal design of WwTP appears to focus in improving the optimization techniques. For instance, a wastewater treatment that includes a tricking filter, an activated sludge aerated reactor and a secondary settler (Mishra, *et al.* 1973) has been repeatedly optimized using different methods: Himmelblau (1976) used an NLP method with a generalized reduced gradient, Mishra, *et al.* (1976) used a simplex pattern search, Casares and Rodriguez (1989) used Integrated controlled random search (ICRS) and Govindarajan, *et al.* (2005) used Adaptive Simulated Annealing (ASA). A wastewater treatment formed by two aerated reactors and a secondary settler has also been optimized with different optimization tools. Gutiérrez and Vega (2002) used the Generalized Benders Decomposition (GBD) algorithm and the same problem was solved by Revollara, *et al.* (2005) using GA. It appears that the trend in the optimization is moving from the use of NLP methods towards the use of random search algorithms, like simulated annealing and GA, perhaps because of the limitations of the NLP with large scale WwTP problems where multi-extrema solutions exists.

Some of the main benefits of optimal design of WwTP as described by the authors presented above are:

- The use of models and optimization methods allows the designer to test different process schemes
- Solutions may include a synthesis of the treatment process and the operational variables (e.g. flow rates or chemical dosing)
- Design solutions found with optimization methods have a lower cost than conventional design methods
- The use of advanced modelling tools allows the evaluation of the design alternatives considering dynamic disturbances (e.g. variation of influent flow and water quality composition).

15

Even though there are significant contributions by researchers in the area of optimum design of WwTP, there are still some gaps and limitations:

- The selection of the treatment processes scheme is normally over simplified. Although, some of the applications found in the literature include variables that allow the optimization algorithm to add or eliminate certain processes, they are based on a pre-selected alternative. For instance, in the application presented by Gutiérrez and Vega (2002), the algorithm proposed can select schemes with one or two aerated reactors.

- The objective function is normally simplified to one single cost function using weighted sums. The optimization of a WwTP design is at least a two objective optimization (i.e. cost and effluent quality), but perhaps because of the mathematical restrictions of the optimization methods (e.g. LP, NLP), most authors include the effluent quality as constraints.

- Even though, the minimization of the cost function is the main objective of the optimization, the cost function itself appears poorly defined. Especially operational costs tend to be overlooked (e.g. include cost of pumping but ignore other operational costs like wages or chemical dosing). In optimization process, better performance indicators could be used when comparing design alternatives as suggested by Vanrolleghem, *et al.* (1996).

- The classical optimization methods applied for solving mixed integer nonlinear problems sometimes fail in presence of discontinuities, get trapped in local minimum and depend strongly on the starting points (Tsai and Chang 2001). Most of the applications have focussed on improving the optimization by using new or different optimization techniques to solve the nonlinear equations with the constraints; but less effort has been put into the use of advanced modelling tools in combination with muti-objective optimization methods.

2.1.3 Optimal waste load allocation

This is a relatively different optimization approach to those presented above for the design of wastewater treatment plants. The waste-load allocation refers to the process of determining the required pollutant removal levels and the location of the discharge points in order to be compliant with water-quality standards in the receiving water body in a cost-efficient manner.

The waste-load allocation problem has been solved using different mathematical programming techniques: LP, SLP, NLP or DP. However, as in the optimization of sewer network or WwTP design, the trend is to use random search techniques. For instance, Burn and Yullanti (2001) use GA to simultaneously minimize the cost of the treatment and pollution impact. As decision variables, the authors consider the removal fractions at each of the point-source locations. The fractions are selected from a set of discrete treatment level options for each point-source location. The GA selects solutions from the available removal fractions and these solutions are evaluated in terms of the cost of the waste water treatment and the water-quality response in the river. The water quality is evaluated as the deviation of the dissolved oxygen (DO) from the standard using a water-quality simulation model implemented in QUAL2E.

In a similar application, Wang and Jamieson (2002) use GA to find the optimal location and level of treatment that minimize the cost and impact on the river measured as the concentration of biochemical oxygen demand (BOD). The impact on the river was evaluated using an Artificial Neural Network (ANN) calibrated using a process based model named

TOMCAT. Cho, *et al.* (2004) solved the location and level of treatment using GA and QUAL2E but these authors considered a wide variety of water quality parameters to assess the impact in the river. They included a composite indicator for DO deficit, BOD, total nitrogen (TN) and total phosphorous (TP) as indicators of the impact on the river. Yandamuri, *et al.*(2006) used the non sorted genetic algorithm (NSGAII) whilst optimizing two objectives: cost and a composite pollution impact indicator based on DO standard violation.

In general, the waste-load allocation problem can be considered as a catchment approach and therefore, it seems to be more holistic. For instance, Zeferino, *et al.* (2009) included in the decision variables, sewer pipes linking urban areas to treatment plants, pumping stations, treatment levels and location of discharge points. However, the interaction between sewer, treatment work and receiving system is highly simplified. In fact, these authors reduce the problem to a single objective, namely the cost, solved using the simulated annealing algorithm.

As this type of problem is considered at the planning level, the components of the UWwSs are normally simplified. No details of the design of the components are included in the optimization. The focus is on the cause-effect relationships between the pollution loads and the water quality impacts in the receiving system (e.g. rivers). Thus, rivers are modelled to some extent. Even though there are a significant number of highly developed models to represent the hydrodynamic and the water quality processes in the rivers, the applications found in the literature use steady state models (QUAL2E) or even a simplified form of the Streeter and Phelps equations. Therefore, the dynamic behaviour of the receiving system is not included. The use of single objective optimization is another common practice; even though most of the latest applications use optimization algorithms that have no restrictions on the number of objectives, the authors limit the analysis to one or maximum two objectives.

Overall, very little interaction between the sewer network, the treatment plant and the receiving system can be found in the literature reviewed. In the following sections some of the applications that included a more integrated view of the system are presented.

2.1.4 Optimal design of the integrated urban wastewater system

In this section a review of research found in the literature that includes the dynamic interaction between components in the design process is presented. As was expected, only a few research publications present any real interaction between components. Table 2.1 compares five characteristics of the research found: the type of problem, the decision variables, the objectives, the integrated model and the optimization method used.

In general, most of the applications are oriented to finding the optimal control strategies for the system under study. It appears that the main driving force of the researchers is to derive control strategies that limit the impact on the receiving system. Examples of this approach are given by Rauch and Harremoës (1999), Schütze, *et al.* (1999) and Meirlaen (2002). As the operation of a treatment plant is driven by the conditions of the influent into the plant from the sewer system, the design of control strategies tends to consider the interaction between the WwTP and the sewer system. An example of this approach is given by Brdys, *et al.* (2008). Only two of the applications found focus on the design of the UWwS components (Langeveld 2004, Muschalla, *et al.* 2006) and only Lobbrecht (1997) presents an example that combines the optimum design and control of the drainage system considering the interaction with the

17

receiving system. However, as can be seen the decision variables are more oriented to general management strategies (storage volume, pumping capacity or infiltration) than for instance the hydraulic design of the pipes in a sewer network or the process design for a WwTP.

Table 2.1 Research examples of design and/or control of integrated urban wastewater systems

Type of application	Decision variables	Objectives	Integrated model	Optimization method	Reference
Design/control	Storage and flow control	Water quality and water quantity	AQUARIUS Urban catchment drainage modelled as non-linear reservoirs. Routing of water quantity and quality	Sequential Linear Programming	(Lobbrecht 1997)
Control	Gates that restrict flow from the urban sub-catchments.	Reduce overflow volumes. River water quality.	Simplified urban runoff process and sewer routing by unit hydrograph. WwTP model using ASM1 River water quality using SAMBA model.	Single objective optimization using genetic algorithm	(Rauch and Harremoës 1999)
Control	Pumping flows Inflow rate WwTP Outflow rate of storm tank Return activated sludge Wasted activated sludge	River water quality	SYNOPSIS: Hydrological and sewer model, ASM1 for the WwTP and a River water quality module. (Later named SIMBA)	Single objective optimization using: Controlled random search, genetic algorithm, gridding, etc.	(Schütze, et al. 2002)
Control	Storm tank Pumping flows Overflow discharge	River water quality	Sewer model in Hydroworks WwTP model with ASM2d River model in ISIS. Surrogate model implemented in WEST	Enumeration of selected alternatives	(Meirlaen 2002)
Control	Storm tanks Pumps Overflows	River water quality	SIMBA5 that include: KOSIM for the sewer, ASM1 for the WwTP and SWMM5 for the river.	Multi-objective Evolutionary Algorithm	(Fu, et al. 2007, Fu, et al. 2008)
Control	WwTP	WwTP Effluent composition	SIMBA: prediction of sewer outflow and model WwTP using ASM2d	Sequential quadratic programming	(Brdys, et al. 2008)
Design	Storage of sewer system and pumping capacity	Investment cost Standards of CSO discharge.	Sewer modelled as storage tank with variable volume. No model of the WwTP or the river was used	Single objective of cost function.	(Langeveld 2004)
Design	Storage volume Throttle discharge Retention soil filters Decentralized infiltration	Investment cost River water quality	Pollution load model WwTP module River water quality module	Multi-objective Evolutionary Algorithm	(Muschalla, et al. 2006)

Integrated design of UWwSs aims to reduce water pollution in the receiving water. This is reflected in the objectives of the applications found in the literature; four out of the six applications reviewed include the so-called river water quality approach as the objective of the optimization. This approach is named water quality or emission based approach

(Meirlaen 2002). The objectives of the other two applications are volume-based and load-based (Langeveld 2004, Brdys, *et al.* 2008) respectively.

One of the main advantages of representing the urban wastewater system as a holistic system lies in the ability to evaluate the performance of the system directly with regard to receiving water quality indicators, rather than by reference to surrogate criteria such as CSO discharge frequency/volume or treatment plant effluent quality (Butler and Schütze 2005). This is of significant importance because of the fact that no close correlations exist between such surrogate criteria and water quality indicators. For example, the reduction of overflow volume is not directly linked to an increase of the oxygen concentration in the receiving water (Rauch and Harremoes, 1999).

In terms of modelling tools, there is a wide variety of software tools used. Four of the applications simulated the three components of the system to some extent. A common approach is to link existing modelling software for each component. It is also common to simplify the model structure of one or two of the components and to consider a more complex structure for the component of interest for the optimization. It appears that the sewer network and the river water quality model is frequently simplified for instance by using hydrological models and tanks in series with variable volume. The simplification strategy seems to be in accordance with the main objectives of the optimization. As the integrated modelling tool is one of the main components of an integrated optimization research, two researchers aimed to develop integrated tools (Meirlaen 2002, Schütze, *et al.* 2002). One of the main limitations encountered by the authors is the long computing times required to evaluate the objective function. The use of surrogate models to replace the computationally demanding process based models appears to be one of the preferred solutions (Meirlaen, *et al.* 2001).

Optimum operation of the urban wastewater system is frequently posed as a mathematical optimization problem and solved by linear programming. But the conceptual restrictions of that technique prohibit the consideration of the complex processes in an integrated UWwS (Harremoës and Rauch 1999). Perhaps this is the reason why the majority of the application used random search optimization techniques. As supported by the research of Schütze, *et al.* (2002) who tested different optimization techniques, the genetic algorithm seems to be the most successful technique. One of the limitations of the research in the optimization of integrated UWwS is the use of single objective optimization, in spite of the fact that in an integrated approach the optimization is by definition a multi-objective problem. The use of multi-objective optimization techniques is one of the recommendations given by Schütze, *et al.* (2002) and successful application of NSGAII are illustrated in the research of Muschalla, *et al.* (2006) and Fu, *et al.* (2008).

In general, three components can be distinguished in the optimization of UWwS. First the problem must be expressed as an optimization problem by defining the decision variables and the objectives of the optimization, second an optimization algorithm to solve the problem must be selected and third the modelling tool to estimate the objectives must be set up. In what follows, the mathematical representation of the general problem of this research is presented and the available optimization methods and modelling tools for UWwS are discussed.

2.2   Mathematical Optimization

2.2.1   Definition of the optimization problem

The design of the component of an UWwS can be posed as an optimization problem in which the aim is to find the combination of decision variables (e.g. pipe diameters, storage volumes or volume of reactors in a treatment plant) which satisfies constraints and optimize the objective of the UWwS (e.g. reduce of flooding or limit the pollution). Hence, optimize means finding such a solution which would give values to all the objective functions that are acceptable to the decision maker.

The decision variables are the numerical quantities for which values are to be chosen in an optimization problem. In an UWwS the decision variables depends on the component that is being designed. For instance, in a sewer network the pipe diameters, slopes and roughness coefficients are common decision variables. As well, for treatment plants, the volume of reactors and the internal flow rates are usually defined as decision variables. Mathematically the decision variables can be represented by a vector x of n decision variables:

$$x = [x1, x2,...., xn] \qquad \text{Eq. 2.1}$$

In most optimization problems there are restrictions imposed by the particular characteristics of the environment or available resources. These restrictions must be satisfied in order to consider a certain solution acceptable. All these restrictions in general are called constraints, and they describe dependences among decision variables and constants (or parameters) involved in the problem (Coello Coello, et al. 2002). In the case of designing a sewer network the restrictions are associated with available commercial diameter of pipes, minimum covering depths of pipes, minimum and maximum flow velocity, etc. These constraints are expressed in the form of mathematical inequalities:

$$gi(x) \leq 0 \quad i = 1,..., m \qquad \text{Eq. 2.2}$$

or equalities:

$$hj(x) = 0 \quad j = 1,...., p \qquad \text{Eq. 2.3}$$

In general the objectives of an UWwS are three folded: to limit the risk of flooding, to reduce the pollution impacts in the receiving system and to reduce the construction and operational cost. Normally these objectives are contradictory. For instance, the aim of the sewer network may be to collect and discharge the storm water generated in the urban area as soon as possible in order to avoid flooding. However, the rapid discharge of peak flows and pollution may affect the water quality and the ecosystem of the receiving system. In addition to being conflicting, these objectives are non-commensurable (measured in different units), so they cannot be compared with each other directly (Fu, et al. 2008) or sum up to form a single objective. Therefore, by definition, the integrated design of an UWwS is a Multi-objective Optimization Problem (MOP). The objective functions are designated as: f1(x), f2(x),...., fk(x), where k is the number of objective functions in the MOP being solved. Therefore, the objective functions form a vector function F(x) which is defined by:

$$F(x) = [f1(x), f2(x),...., fk(x)] \qquad \text{Eq. 2.4}$$

A general MOP is defined as minimizing (or maximizing) F(x) = (f1(x),...., fk(x)) subject to the inequality constraints gi(x) ≤ 0, i = {1, ..., m}, and equality constraints hj(x) = 0, j =. A MOP solution x minimizes (or maximizes) the components of a vector F(x), where x is a n-dimensional decision variable vector x = (x1,...., xn) from some universe Ω (Coello Coello, *et al.* 2002).

The set of all n-tuples of real numbers denoted by R^n is called Euclidean n-space. Two Euclidean spaces are considered in MOPs:
- The n-dimensional space of the decision variables in which each coordinate axis corresponds to a component of vector x.
- The k-dimensional space of the objective functions in which each coordinate axis corresponds to a component vector $f_k(x)$.

Every point in the first space represents a solution and gives a certain point in the second space, which determines a quality of this solution in terms of the objective function values.

2.2.2   Definition of Pareto terminology

Having several objective functions, the notion of "optimum" changes, because in MOPs, the aim is to find good compromises (or "trade-offs") rather than a single solution as in global optimization. The notion of "optimum" most commonly adopted is that originally proposed by Vilfredo Pareto. The definition says that:

> "x* is Pareto optimal if there exists no feasible vector x which would decrease some objective function without causing a simultaneous increase in at least one other objective (assuming minimization)."

Additionally, there are a few more definitions that are adopted from Coello Coello, *et al.* (2002):
- Pareto optimal set (P*): are those solutions within the genotype search space (decision variables space) whose corresponding phenotype (objective vector) components cannot be all simultaneously improved. These solutions are also termed *non-inferior*.
- Pareto front set (PF*): are the vectors in the objective function space that correspond to the evaluation of the Pareto optimal set of solutions (P*). These vectors are also named *nondominated*. When plotted in objective space, the nondominated vectors are collectively known as the Pareto front.
- Pcurrent (t): at any given generation of a MOP, a "current" set of Pareto solutions (with respect to the current generational population) exists and is termed Pcurrent (t), where t represents the current generation number.
- Pknown (t): is a secondary population which is referred to as an archive or an external archive, in order to store non-dominated solutions found through the generational process.
- Ptrue: the true Pareto solution set (termed P*) is defined in the computational domain as Ptrue which is usually a subset of P*. Ptrue is defined by the functions composing a MOP and the given computational domain limits.
- PFcurrent (t), PFknown (t), and PFtrue: are the Pareto front sets corresponding to the Pcurrent (t), Pknown (t), and Ptrue sets.
- Acceptable compromise solution: The decision maker typically chooses only a few points in PFknown as generated by Pknown. The associated Pareto Optimal solutions are then the "acceptable" (by the decision maker) compromise solutions. The decision makers base their solution choice taking into account the non-modelled human's preference.

2.2.3   Overview of optimization methods

This section provides a review of various optimization techniques described in the literature. The discussion will focus in those techniques that can solve MOPs and the aim of this section is to support the decision on which type of optimization method is more appropriated for use within this research.

Available methods for MOP can be classified in different ways. One of them is based on whether many Pareto-optimal solutions are generated or not, and the role of the decision maker (DM) in solving the MOP. This particular classification was adopted by Diwekar (2003). The DM is in charge of selecting one of the Pareto-optimal solutions for implementation based on their experience and other considerations not included in the MOP. Multi objective optimization methods are divided into two main groups: generating methods and preference-based methods. As the names imply, the former methods generate one or more Pareto-optimal solutions without any inputs from the DM. The solutions obtained are then provided to the DM for selection. On the other hand, preference-based methods utilize the preferences specified by the DM at some stage(s) in solving the MOP. Preference-based methods require preferences in advance from the DM or they may play an active role during the solution by interactive methods (Rangaian 2009). For the MOP study in this research is difficult to specify preferences without or with limited knowledge on the optimal objective values and is not practical at this stage to continuously involve a DM. Therefore, for this research generating methods are preferred. They provide many Pareto-optimal solutions and thus more information useful for decision making is available. The role of the DM is after finding optimal solutions, to review and select one of them.

Another classification is based on the method of operation. In this classification, the optimization techniques are divided into two categories: deterministic and stochastic (random). Deterministic algorithms do not contain instructions that use random numbers in order to decide what to do or how to modify data. On the contrary stochastic algorithm includes at least one instruction that acts on the basis of random numbers. Stochastic algorithms are also often called randomized or probabilistic algorithms. Figure 2.1 shows the classification of the global optimization methods and some examples of each category.

Enumerative algorithms use an exhaustive evaluation of all possible alternatives. Considering that the size of the search space in the optimization of a water system is huge, enumerative approach seems to be no feasible for practical applications. Such an approach would take an infeasible long time.

Deterministic algorithms will always produce the same results when given the same inputs. For many problems however, deterministic algorithms are unfeasible. In global optimization, the problem space is often extremely large and the relation of an element's structure and its utility as solution is not obvious (Weise 2009). In the optimization of UWwS the relation between an alternative solution and its objective functions is very complex and the dimensionality of the search space is very high. Therefore to solve the MOP deterministically is highly complicated. Trying it would possibly result in exhaustive enumeration of the search space, which is not feasible even for relatively small problems.

Greedy and hill-climbing algorithms, branch and bound tree/graph search techniques, depth- and breadth-first search, best-first search, and calculus based methods are all deterministic methods successfully used in solving a wide variety of problems. However, many MOPs are high dimensional, discontinuous, multimodal, and/or nondeterministic polynomial time (NP-

Complete). For instance, the optimum design of storm drainage has proved to be a complex NP optimization problem, encompassing multimodal, discontinuous (or mixed discrete-continuous), non-convex features (Guo, *et al.* 2008). Deterministic methods are often ineffective when applied to NP-Complete or other high dimensional problems because they are handicapped by their requirement for problem domain knowledge (heuristics) to direct or limit search in these exceptionally large search spaces. Problems exhibiting one or more of these above characteristics are termed irregular (Coello Coello, *et al.* 2002).

MOP complexity and the shortcomings of deterministic search methods promote the creation of several optimization techniques by the operations research community. These methods (whether linear or nonlinear, deterministic or stochastic) are normally grouped under mathematical programming methods. Linear programming is designed to solve problems in which the objective function and all constraint relations are linear. Conversely, nonlinear programming techniques solve some MOPs not meeting those restrictions but require convex constraint functions. It is noted here that many problem domain assumptions must be satisfied when using linear programming and that many real-world scientific and engineering problems may only be modelled by nonlinear functions (Coello Coello, *et al.* 2002). Finally, stochastic programming is used when random-valued parameters and objective functions subject to statistical perturbations are part of the problem formulation. Depending on the type of variables used in the problem, several variants of these methods exist (i.e., discrete, integer, binary, and mixed-integer programming). As noticed by Schütze, *et al.* (2002), these methods require making specific assumptions on the optimization problem that they appear not to be appropriate for the optimization of UWwS.

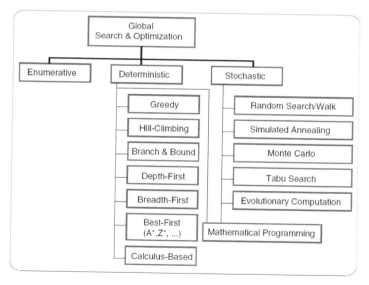

Figure 2.1 Global optimization methods

Source: (Coello Coello, *et al.* 2002)

As an alternative to solve irregular problems, different stochastic methods have been developed: Simulated Annealing (SA) (Kirkpatrick, *et al.* 1983), Monte Carlo methods (Osyczka 1985), Tabu search (Glover and Laguna 1997), and Evolutionary Computation (EC)

(Goldberg 1989). Stochastic methods require a function assigning fitness values to possible (or partial) solutions, and an encode/decode (mapping) mechanism between the problem and algorithm domains. Although some are shown to "eventually" find an optimum most cannot guarantee the optimal solution. They in general provide good solutions to a wide range of optimization problems which traditional deterministic search methods find difficult.

A random search is the simplest stochastic search strategy, as it simply evaluates a given number of randomly selected solutions. Like enumeration, though, these strategies are not efficient for many MOPs because of their failure to incorporate problem domain knowledge. Random searches can generally expect to do no better than enumerative ones. In general, Monte Carlo methods employs a pure random search where any selected trial solution is fully independent of any previous choice and its outcome (Osyczka 1985). The current "best" solution and associated decision variables are stored as a comparator. Tabu search is a meta-strategy developed to avoid getting "stuck" on local optima. It keeps a record of both visited solutions and the "paths" which reached them in different "memories." This information restricts the choice of solutions to evaluate next. Tabu search is often integrated with other optimization methods (Glover and Laguna 1997).

SA is an algorithm developed based on the analogy with the thermodynamics process "simulated annealing" where crystals attain their state of minimum energy best by being cooled down slowly. SA chooses the best move at random occasionally allowing moves in different directions. If the move improves the current optimum it is always executed, else it is made with some probability $p < 1$. This probability exponentially decreases either by time or with the amount by which the current optimum is worsened (Kirkpatrick, et al. 1983). The analogy for SA is that if the "move" probability decreases slowly enough the global optimum can be found.

EC is a generic term for several stochastic search methods which computationally simulate the natural evolutionary process. EC embodies the techniques of genetic algorithms (GAs), evolution strategies (ESs), and evolutionary programming (EP), collectively known as Evolutionary Algorithms (EAs). These techniques are based on natural evolution and the Darwinian concept of "Survival of the Fittest" (Goldberg 1989). Common between them are the reproduction, random variation, competition, and selection of contending individuals within some population. In general, an EA consists of a population of encoded solutions (individuals) manipulated by a set of operators and evaluated by some fitness function.

Evolutionary algorithms seem particularly suitable to solve MOPs, because they deal simultaneously with a set of possible solutions (the so-called population). This allows finding several members of the Pareto optimal set in a single run of the algorithm, instead of having to perform a series of separate runs as in the case of the traditional mathematical programming techniques. Additionally, evolutionary algorithms are less susceptible to the shape or continuity of the Pareto front, whereas these two issues are a real concern for mathematical programming techniques. Some MOP solution approaches focus on search and others on multi-criteria decision making. Multi-Objective Evolutionary Algorithms (MOEAs) are then very attractive MOP solution techniques because they address both search and multiobjective decision making. Additionally, they have the ability to search partially ordered spaces for several alternative trade-offs (Coello Coello, et al. 2002).

Generally speaking, the objectives are non-commensurable, so they cannot be compared with each other directly. Definitions of the objective functions usually involve water flow and quality states in the system, so their evaluations can only be achieved through use of an integrated model. A simplified representation of the objective functions (for example, linearization) seems to be problematic for the integrated system due to its complexity (Schütze, *et al.* 2002). In this situation, stochastic optimization methods are most appropriate and therefore evolutionary algorithms are considered the best alternative for this research.

2.3 Modelling Tools for Urban Wastewater System Components

2.3.1 Modelling the components of the urban wastewater system

In this section are described the software tools typically used to model the fundamental mechanisms and processes in the each component of the UWwS. A detailed review of those software's is presented by Freni et al (2003) and Rauch et al (2002), however here the emphasis is on the integrated software options and their limitations. Table 2.2 shows a summary of some of the software available for modelling the components of the urban wastewater systems. The list was updated based on the table presented by Price (2000) and the software analysed by Freni et al (2003). Notice that the list of modelling tools does not pretend to be fully complete, and may only present models reported in scientific literature and may be available for use by public in general.

Table 2.2. Software for different components of the urban wastewater systems

Sewer network		Wastewater Treatment Plant		Water receiving system - River	
HydroWorks (InfoWorks)	Wallingford Software	GPSX	Hydromantis	MIKE 11	DHI
MOUSE (Trap)	DHI	STOAT	WRc	ISIS	HR Wallingford/ Halcrow
SOBEK- Urban	Delft Hydraulics	EFOR	Kruger	SOBEK	Delft Hydraulics
CANOE[1]	Sogreah	SIMBAD	CGE	SIMPOL	WRc
SIMPOL	WRc	WEST++	University of Ghent	NWMB	MUNW
MicroDrainage	MicroDrainage	SIMBA	ifak	DMZ	RIZA
SWMM	US-EPA[2]	BIOWIN	EmviroSim Associates Ltda	MCARLO	UKEA
ILSAX	University of Technology, Sydney	AQUASIM	Swiss Federal Institute for Environmental Science and Technology	STREAMIX	US-EPA
KOSIM	Institut fur technische-wissenschaftliche Hydrologie, Hannover			DUFLOW	STOWA
STORMCAD/ SEWERCAD	Haestad Method			WASP	USEPA
COSSMOS	University of Reggio Calabria (Italy)			QUAL2E	US-EPA
DORA/DORAT	Ars Nova Multimedia			MIKE-SHE	DHI
				SIMCAT	TW
				River Water Quality Model No1	IWA

[1]. Does not model water quality – yet. [2]. Front-ended by a number of organisations. Source: adapted (Price 2000).

The availability of models is quite large and their characteristics are widely variable, ranging from very simple conceptual lumped models to complex process-based distributed models. Their development follows the historical separated development of the UWwS itself. The sewer models are closer to the river models, with a strong development in the hydrodynamic process, with less development in the water quality issues, especially in the sewer networks. In contrast, the WwTP models are very well developed in water quality but use a simplified representation of the system hydrodynamics.

2.3.2   Integrated modelling software.

The first idea to integrated modelling was made by Beck in 1976, but it took until middle of 1990s to include deterministic models of the total system. This happened when the technical understanding of the sub-systems, the computer simulation capabilities, and the institutional frameworks had matured sufficiently to allow the first steps to be made towards producing viable integrated representations of the full urban wastewater system (Rauch, *et al.* 2002, Butler and Schütze 2005).

The integrated modelling allows various design and operational scenarios and their impacts on the environment to be studied without having to alter physically the system or to set up physical laboratory-scale models. Thus, a substantial amount of financial expenditure and effort can be saved by computer simulation. Representing and understanding the urban wastewater system as a whole allows better, more cost-effective solutions to be engineered because consideration of just the individual elements does not take into account the subtle interactions between the various subsystems (Butler and Schütze 2005).

The models presented in Table 2.2 were developed independently for the different parts of urban wastewater systems. Consequently, the numerical models as well as the simulation software were lacking direct interfaces. Recognising the necessity of the integrated approach, serious efforts were made to combine existing software packages. The characteristics of some of the software available for integrated urban wastewater system modelling are presented in the Table 2.3.

2.3.3   Limitations in the integration of models

The integrated modelling of UWwS inherited the same fragmentation in components as the development of the wastewater system itself. Detailed process-based models of sewer, treatment plants and receiving water bodies were created to describe the performance of each subsystem (Rauch, *et al.* 2002). The most common approach to develop integrated UWwS models is based on the combination of existing models of each subsystem. However, this approach brings consequently the following limitations: different state variables, limited flux of information between models of subsystems and long calculation times.

Models with different state variables
The models to be merged have incompatibilities between state variables, parameters and processes. An example of different state variables is given by the WwTP model ASM that is based in COD and the river models like QUAL2E, MIKE11 or ISIS that are based on BOD (Meirlaen 2002). In conventional sewer models, pollutant concentrations are frequently assumed equal in different events and constant during the event. Thus, conversion processes are neglected. If conversion processes are included, they correspond to those developed for rivers, based on BOD. Therefore, the conversion from particulate and dissolved BOD in sewer to the COD fractions in ASM and back to BOD of rivers models is a key problem that limits the reliability of present integrated models (Rauch, *et al.* 2002).

Table 2.3. Characteristics of the software available for integrated urban wastewater system modelling

Characteristics / Software Name	ICS	SYNOPSIS	SIMBA	WEST	CITY DRAIN
Developer	DHI – Horsholm, DK and WRc, Swindon, GB.	Imperial College, UK	Ifak, Berleben Germany.	Hemmis NV. Ghent University, Belgiun.	Achleitner et al, 2006. University of Innsbruck, Austria.
Year	1999	1998	2002	2002	2005
Models merged: Sewer system	MOUSE	KOSIM (cascades of linear reservoirs) Limited Backwater S	PLASKI + SIMBA Sewer	KOSIM	Modified MUSKINGUM (conceptual model)
Models merged: WWTP	STOAT	Simplified ASM1	SIMBA (biological and chemical treatment + sedimentation)	WEST (CSTRs)	Black Box model (Pollutant removal percentages) and ASM1
Models merged: Receiving System	MIKE 11	DUFLOW	SIMBA Sewer	RWQM1 (CSTRs model)	MUSKINGUM (conceptual model)
Water Quality Modelling	Transport, conversion process of water and pollutants	No biochemical transformation in Sewer system	Transport and conversion process of water and pollutants	Transport and conversion process of water and pollutants	Mass transport of conservative matter. No transformation process
Control Module	Yes	Yes	Yes	Yes	
Optimization Module		Yes	Yes	Yes	
Platform			Open platform based MATLAB/ SIMULINK	Open Simulator using MSL-USER	Open platform based MATLAB/ SIMULINK
Bidirectional interaction Additional modules between models	Yes	No	Yes	Yes	Yes
Truly synchronised simulation	Yes	Only Sewer and WWTP	Yes	Yes	Yes
Simulation of control options possible	Under development	Only Sewer and WWTP	Yes	Yes	Yes
Simulation of long time series feasible	Under development	Yes	Under development	Yes	Yes
Open simulation environment	No	No	Yes	Yes	Yes
Integrated use at a real case study reported	Yes	Semi hypothetical	Yes	Semi hypothetical	Yes

(Adapted from Rauch et al (2002) and Meirlaen (2002))

Two basic approaches have been applied to deal with the problem. One approach is to develop a *completely new COD based model* as the RWQM1 (Somlyody, *et al.* 1998), even though it still has some incompatibilities. Other approaches *use connectors* that include a logical transformation of the state variables and have a closed mass and elemental balance. This pragmatic approach has been implemented for instance in the WEST platform. The problem with the last approach is that it uses fixed relationships between variables and it is known that, for example, COD relations can change during a storm or be different between storms (Meirlaen 2002). Even more, the conversion factors depend on the characteristic of the sewer system, on the evolution of the specific event and on the dynamics of the compound transport. Therefore, according with Rauch et al (2002), one of the major requirements for the further development of integrated modelling is to develop consistent sets of state variables in the subsystem models in order to be able to run them without any conversion factors at the interface of the models.

Fluxes of information between different parts of the model
There are two approaches to develop an integrated model the *sequential approach* and the *simultaneous approach*. The sequential approach implies the use of three models that run one after the other over the whole simulation period, using the output of one model to feed the next model. In this approach, the fluxes proceed in the forward direction. On the contrary, in the simultaneous approach all the elements of the system are computed simultaneously and the fluxes of information can go forward and backward (Meirlaen 2002).

It is not always needed to have simultaneous simulation. For example, the sewer system could be considered as a process that proceeds only in a forward direction. However, when feedback fluxes appear like the return of sludge in a WWTP, the process is no longer in the forward direction. In the design of integrated real time control (RTC) strategies for UWwS, the fluxes of water quantity, water quality compounds and control signals in both directions have to be considered (Schütze, *et al.* 2002). Thus, usually in this type of application, the complex simultaneous simulations are needed in order to take into account the system dynamics and to design, tune and implement RTC strategies (Rauch, *et al.* 2002).

The simultaneous simulation depends on the approach of the software implementation for the integrated model. The first and more common approach is to merge different software tools. In this approach having a simultaneous simulation requires a coordinating program in order to exchange information (either directly or via a file) between the different software tools used. This approach is implemented for example by SYNOPSIS (Schütze 1998) and for the Integrated Catchment Simulator ICS (Clifforde, *et al.* 1999). The second approach is to use a common simulation platform where the integrated model is created by assembling a set of elements (pipes, structures, basins, river reaches, etc). This approach has been implemented in open simulators as SIMBA and WEST (Rauch, *et al.* 2002).

According to Meirland (2002), the coordinating program induces an overhead and the use of conversion factors for different estate variables cannot be avoided. On the other hand, the implementation of all components in one package as with the WEST simulator is still inefficient increasing the computational time to have simultaneous simulation. It should be evaluated which of the existing implementations is better in a given situation.

Long calculation times
Even though the limitation of fluxes of information has been partially solved with the approaches mentioned above, the issue remains that simultaneous simulations of integrated UWwS require long calculation times. The problem becomes more complex as explained by Schütze, *et al.* (2004a) when the "*model based predicting control*" approach is used to develop the control procedure in the system. This approach consists in setting up an on-line model that at every time step evaluates the impact of a control action on the system and applies the one that is more beneficial for it based usually on the optimization algorithms. Depending on the complexity of the model the calculation time can be a critical issue, since a

potentially large number of different control actions and their impacts on the wastewater system will have to be evaluated within a short time.

Another application in which the long calculation times could be a limiting factor is in *"system design through rigorous optimisation"*. Many design problems might be solved more straightforwardly by applying an analytic design procedure, using (mathematical) multi-criteria optimisation. Due to the complexity and size of integrated UWwS, mechanistic models are not suited for direct system design using optimisation; rather; they are most often used to check designs. Systematic system design employing multi-objective optimisation would be facilitated if simple models containing the most important phenomena were available (Weijers 2000).

The need for *simplified models* in wastewater engineering has resulted in a variety of reduction approaches summarized as follows according to Weijers (2000).
- *New model building from scratch.* Even though it is not a straightforward reduction, most of the time, explicitly or implicitly the knowledge of mechanistic models is included in the new model developed.
- *Simplifying assumptions.* This may include simplification of components (e.g. aggregation of variables), processes, (e.g. aggregation of reactions), lumping of space distribution, and kinetics (e.g. simplification of complicated kinetic schemes).
- *Neglect of dynamics by quasi-steady-state assumptions and singular perturbation.* The control problem is decomposed into a hierarchical set of several levels of smaller sub-problems. On each level, the dynamics of lower levels are assumed to be very fast and considered to be in (pseudo)-steady-state and the dynamics of higher levels are assumed to be very slow and considered as constant. Singular perturbation is a mathematical technique to analyse timescale multiplicity and to perform a systematic order reduction and error analysis.
- *Order reduction methods.* In modern control engineering, order reduction of linear models is a very important task in control system design. Models for control are often obtained from the linearization of high dimensional rigorous models, resulting in very high dimensional models. Modern controller design methods, especially robust control design methods, yield even higher dimensional controllers. Order reduction of a model or controller has become a necessity if it is to be implemented, as high-order controllers are usually not accepted in industry.
- *Black–box identification.* To construct reduced models of nonlinear systems with validity over a larger domain, nonlinear black box modelling techniques may be applied (e.g. artificial neural networks (ANN)).

Vanrolleghem *et al* (2005) presents the two main ideas that have been implemented to reduce the calculation times in integrated UWwS modelling: the first is *model simplification* with the use of mechanistic surrogate models, and the second is *model reduction through system and time boundary relocation*.

The model simplification: is based in the "replacement" of the complex mechanistic models (e.g. those which use Saint Venant equations) with a more simple model (e.g for those that use Combined Stirring Tanks Reactors CSTR), which is faster and less accurate but still sufficiently for the purpose needed. The models obtained with the approaches mentioned above differ in their degree of "greyness". Models that are obtained via systematic reduction of a "white" mechanistic model whilst preserving the physical interpretation of the system states can be considered light grey. Simple, mechanistic input-output models are considered grey. The last category is black box models, such as polynomial models or artificial neural networks. It is noted that also mixed forms can be applied (Weijers 2000). Examples of simplified models are the conceptual sewer model Kosim that is used in Germany as a tool for the design of sewer systems (Butler and Schütze 2005) and was implemented in SYNOPSIS and WEST to describe the sewer system, or the conceptual flood routing model Muskingum used in CITYDRAIN to simulate a sewer network (Achleitner, et al. 2007).

One of the problems with the model simplification approach is that in those models less information is contained in the equations as compared with Mechanistic (process-based) models. Thus, more data are needed to compensate for the lack of information from the knowledge side. Many parameters no longer have a strict physical meaning, as they are the result of "lumping". The monitoring campaign to collect data to calibrate a simplified model could be rather expensive (Vanrolleghem, et al. 2005). Then, instead of trying to collect all the data from reality, generated "data", using complex mechanistic models is used to calibrate simplified models. Meirlaen et al (2001), describe the methodology used to calibrate an integrated UWwS model using the WEST platform. However, this methodology implies an enormous effort in terms of calibration and verification of two models for the same system in order to tackle the long calculation time problem.

The model reduction through system and time boundary relocation: Model reduction for sewers and rivers focuses on the fact that in some cases, parts of the system do not need to be modelled but can be replaced by boundary conditions. This is especially useful in the case of integrated RTC strategy design. For the design and tuning of RTC strategies Meirlaen (2002) identified four types of possible model reductions: relocation of upstream and downstream boundaries, reduction of the conversion model and reduction of the time boundaries. All these reductions are based on the facts that the controller under study is only influenced by certain parts of the system and only influences part of the system, both in time and space. Only those parts need to be modelled when designing and tuning the control strategy. This approach is applied in integrated urban wastewater modelling for example by Lobbrecht (1997), Meirlaen (2002) and Erbe et al (2002).

Another aspect strongly influencing the calculation time is the software implementations and numerical algorithms used for solving the model equations. It is clear that compiled code is superior to interpreted code (with respect to calculation time). Most integration and optimisation algorithms currently available in commercial software are rather old, and new techniques, like genetic algorithms, are only slowly being implemented. Also efficient stiff solvers are not implemented currently in WwTP models, which limits their performance for the integration of stiff model equations, with the biological models used (Meirlaen 2002).

Development of properly integrated software
According to Price (2000) one factor that affects the development of properly integrated software is that there is no demand for them. Despite the public interest in the impact on the environment and the efforts being made by the wastewater managers, the commercial demand is still limited. An important explanation to why there is still not a wide application of integrated models could be the split of responsibilities for the management and planning of sewers, treatment plants and receiving systems (Rauch, et al. 2002). Hence, the development of integrated software is hampered also by administrative fragmentation.

In 2000, Price suggested that we have yet to see a formal commercial product that is integrated at each level: database, model building, process interpretation, etc and ten years later still we are expecting it in spite of the actual software development. Perhaps, because of the present wide range of views of the different phases and how they should be modelled, the need for better scientific knowledge on the processes involved, and the weakness of commercial demand, a truly integrated model addressing the *detailed* physical, chemical and biological aspects that is accepted internationally may not emerge for several years. In the mean time we will be making more extensive use of much simpler, integrated models and making better use of sensitivity and uncertainty analysis to overcome the considerable deficiencies of these models (Price 2000). However, in the last years, efforts have been devoted to make them accessible and sufficiently performing for practical development of solutions for urban wastewater systems (Schütze, *et al.* 2003). Their simplicity enable us to do time series analysis and to gain an understanding of how the integrated system works and its operation improved (even in real time) (Price 2000).

In conclusion, the new tools allow us to do analysis, design, operation and real time control of the integrated urban wastewater systems and some implementations have been recorded, but there appears to be a significant conservatism in the profession with respect to the application of new tools that still is limiting the application of new technologies.

2.4   Conclusions

The literature reviewed show that the optimum design of the urban wastewater considering the interaction between components is a research field in a very early stage. Event though there are some significant experience in the optimum design of each component separately, the research presented here is relevant. Some of the gaps that this research covers are:

- The definition of a methodology for the design of an UWwS component considering the interaction between components. The main methodological contribution so far can be attributed to the work of Schütze, *et al.*, (2002) but is mainly for the development of integrated control strategies.
- The case studies presented in the literature that consider a holistic approach are mainly for the definition of control strategies. The few examples of design found in the literature are insufficient to demonstrate the potential of optimum design of UWwS. They are over simplified in terms of the case study and the level of complexity used to represent the system.
- The benefits and limitation of the integrated applications are not clearly identified.

In terms of the components of the approach, the reviewed literature also allows us to make some important decisions for the research:

- The problems of interest are Multi-objective Optimization Problems (MOPs)
- The trend in the optimization techniques used to solve the MOPs pointed to random search algorithms. Within those available in this category, the MOEAs appear to be the best alternative for this research.
- The integrated modelling tools available appear not to be appropriate for the research in hands. This implies that for the development of the case studies, the approach to follow will include the development of an integrated modelling tool by linking existing software for each component. The level of complexity depends significantly on the objective of the optimization; therefore the level of complexity may be defined in the course of the development of the case studies.

In the following chapter a general methodology for this research is described. The concepts and definitions presented in this chapter are used to craft the method required for optimum design and control of UWwSs.

3 Framework for Optimum Design and Control of Urban Wastewater Systems

3.1 Introduction

This chapter describes the approach proposed for this research in order to achieve its objective: the optimum design and control of Urban Wastewater Systems (UWwS). The approach is presented here to maintain the logical description of the document (introduction, literature review, methodology, results and conclusions). However, the approach proposed has been influenced by the experience gained in the realization of the case studies and incorporates the lessons learned using the approach. The scope of this research is to contribute to the reduction of urban pollution affecting receiving water systems through the optimization of the design and control of the integrated UWwSs. The challenge of the research is to develop a design approach that includes the interaction between the components of the UWwS; the sewer, the wastewater treatment plant (WwTP) and the river. In addition, the approach should have the right balance between the traditional design approaches and the use of advance modelling tools and optimization algorithms.

3.2 Conceptual Framework for Model Based Design and Control

The Model Based Design and Control (MoDeCo) approach can be described as a combination of the iterative design and model predictive design approaches, as defined by Harremoës and Rauch (1999) and described in Chapter 1. Thus, MoDeCo approach starts with a pre-design that is based on traditional approaches and empirical rules of UWwS operation. The pre-design is used to build the model of the system (e.g. sewer), together with information from the other components of the system (i.e. WwTP and the river). Then, the model plays the role of representing the real UWwS and is used to predict the performance of the system. Alternative solutions are automatically generated and a set of optimum solutions is found using multi-objective optimization algorithms. The first draft of the conceptual framework was presented in Vélez, *et al.* (2007). The final conceptual framework for model based design and control includes six steps: problem definition, design of UWwS components, design of operational strategies, implementation of the model of the system, optimization and post-processing. The flow chart for MoDeCo is presented in Figure 3.1. Each of the steps in the approach is presented in the following sections.

3.2.1 Identify the problem

The first step in the approach is to identify the problem to be addressed. In general, three cases of UWwS design can be expected:

i. All components of the UWwS exist and therefore the optimisation of the system is based on the design of control strategies (i.e. optimum operation);
ii. One component of the system does not exist, thus the optimisation will include the design of the missing component and the operational strategy
iii. The UWwS is completely new, and therefore the degree of freedom is full and the optimisation problem becomes the most complex possible.

In the last case, the full MoDeCo approach (as presented in Figure 3.1) can be applied to optimise the UWwS.

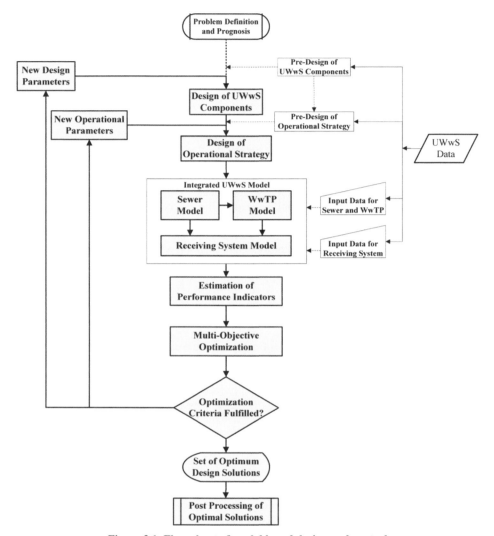

Figure 3.1 Flow chart of model based design and control

Following the identification of the problem, the objectives of the UWwS must be defined. In general, the objectives may be associated to water quantity, water quality and cost. The objective can be assessed using UWwS performance indicators (PIs). Examples of PIs can be found in (Matos, *et al.* (2003). Considering the problem and objectives defined, an appropriated technology should be selected for the solution of the problem. For instance, choose between combined or separated sewer networks when designing a drainage system or between activated sludge or stabilization ponds when designing a wastewater treatment facility. These steps follow traditional design approaches; therefore not much detail is included in this section on the diagnosis of the problem and the selection of the technology.

3.2.2 Design the urban wastewater system components

The design process requires the data from the urban catchment and a preliminary design of its components according to the problem being addressed. The preliminary design is based on traditional approaches, considering steady state conditions and empirical rules. General approaches for preliminary designs can be found elsewhere. For instance, for urban drainage Butler and Davies (2000) can be consulted, and for the treatment plant design the wastewater engineering book of Tchobanoglous, *et al.* (2003).

The design rules have to comply with local regulations and conditions and are highly dependent on the specific case. In general, the preliminary design is considered in MoDeCo as a sub-optimal solution of the problem. One step further in MoDeCo approach is the definition of the decision variables based on the preliminary design. These are the parameters used in the optimization routine to generate new design solutions. In the process they are generated by the optimization algorithm and become an input for the design (Figure 3.1).

3.2.3 Design of the control strategy

If the problem includes the design of the control strategy, then the design should fulfill the operational objectives and performance indicators as defined in the first step. In general, the control design includes the definition of the actuators (e.g. pumps, weirs, gates, etc), the definition of information used to operate the actuators (e.g. water levels, flows, etc) and the set points. Preliminary design of the control strategy can be achieved following the methodology described by EPA (2006) and Meirlaen (2002) or Schütze, *et al.* (2002). For MoDeCo approach, the predesigned control strategy is parameterized (e.g. vector of set points) and those parameters are used in the optimization algorithm to generate new control strategies. New parameters become an input to the design process of the control strategies.

3.2.4 Implement the model of the system

The model of the wastewater system provides a quantitative cause-effect relationship between the design parameters and the performance indicators. The Model is used here both as a mathematical representation of the designed system (process model) and for predicting system behaviour under various operational strategies (control model). The model complexity is highly depending on the specific case being designed and the objectives. Some general rules to define the complexity of the integrated model are proposed in the UPM manual (FWR 1994), while Rauch, *et al.* (1998a) propose the selection of the model based on the water quality impacts in the receiving system. In general, good modelling practices should be follow as those presented by Van Waveren, *et al.* (2000) and (Rietveld, *et al.* (2010). As it was shown in Chapter 2, the existing integrated modelling tools have limitations; and therefore, to achieve the objectives of this research a new integrated modelling tool was developed. The integrated model developed for MoDeCo is presented in section 3.4.

3.2.5 Generate new design alternatives and select the best solutions

The generation of new design and operational alternatives, as well as the selection of the best solutions is the task of the optimization algorithm. The optimization process is the core of MoDeCo. Here, the PIs of the system generated by the UWwS model are used to assess the competing solutions of the design. Using different meta-heuristic approaches, the optimization algorithm can automatically create new design and control parameters that will form new hydraulic and process design and/or control strategies. The optimization algorithm loops until it fulfills stopping criteria (e.g. convergence criteria or number of solutions evaluated). As stated in Chapter 2, this research addresses mainly multi-objective problems

(MOPs) and one of the best alternatives to solve this type of problems is using multi-objective evolutionary algorithms (MOEAs). The MOEA implemented in MoDeCo is described in section 3.5.

3.2.6 Post-process optimum solutions

Different to the single objective optimization, the solution of MOPs is a set of solutions which require further analysis. In this case, a post-processing is required in which the preferences of the decision makers are included. For the sake of simplicity, in this research we select few alternatives from the group of solutions for further analysis considering extremes situations (for instance, the cheapest and the most expensive solution found with the algorithm). A more detailed analysis includes the preferences of the decision makers, the evaluation of the alternatives with long term continuous simulation and the estimation of long term PIs.

3.3   Data Requirements

In general, there are two types of data required: i) the data that describe the infrastructure (buildings, roads, pipes, storage tanks, treatment reactors) and ii) and the data that describe the water that pass through that infrastructure (precipitation, water levels, flows and water quality components). Data needed to design an UWwS using MoDeCo approach are influenced by the following aspects:
- The type and complexity of the problem being solved.
- The modeling approach selected to represent the system.
- The performance indicators selected to compare the design alternatives
- The cost of data acquisition.

For instance, if the problem is the design of the components of the UWwS (sewer and WwTP) there is no data available to calibrate and validate models except for the model of the receiving system for which data should be collected. If the problem is the design of the control strategy of an existing system, data for calibration and validation of the models should be collected. Figure 3.2 shows a scheme with the data needed for the design of an UWwS using the MoDeCo approach. The scheme represents a combined sewer system with storage and treatment plant discharging into a river.

Without any doubt the data that describe the infrastructure is of fundamental importance and the accuracy in the description of the infrastructure will influence the second type of data: the water quantity and quality generated in the urban wastewater system. For instance, Clemens (2001) found that limitations in the structures and geometry data used to set up a drainage model may produce errors with the same order of magnitude as measuring inaccuracies. But, perhaps is the water quantity and quality, the less available information to build the models. For instance, the sizes of the pipes in drainage depend on the design rainfall and the rainfall is the major disturbance in the operation of an urban wastewater system. However, is common to find rainfall data with limited spatial and temporal resolution. In sewer networks the water-quantity data is scarce and the water-quality data is more limited. Even more, the data available in sewer systems often has low sampling frequency (Winkler, *et al.* 2004). This may limit any analysis that addresses transient pollution events.

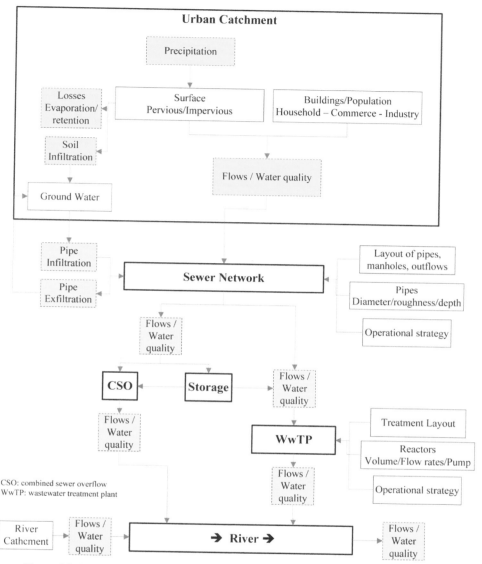

Figure 3.2 Scheme of data requirements for the design of an urban wastewater system

The availability of data from WwTP may be better than the other two components of the UWwS (i.e. sewer and receiving system). This may be explained by the fact that treatment processes are concentrated in one location which facilitates the implementation of, for instance, automatic monitoring networks. However, when the data is collected for operational purposes, it may lack the sufficient spatial and temporal resolution to identify effects of precipitation events inside the process (i.e. normally inflows and outflows are monitored). Rivers normally have better control of the water quantity. However, data of water quality is limited, especially for those parameters that may be impacted by discharges from sewers and WwTPs. Even though there are a significant effort in improving the information in urban areas, for instance, with the implementation of monitoring networks for WFD (EC 2003), set of data for integrated modelling of UWwS is scarce. The lack of data is perhaps one of the major limitations for the design of UWwS considering the dynamic interaction between components (Langeveld, *et al.* 2003).

3.4   Integrated Modelling Tool of Urban Wastewater System

The main objective of the integrated modelling tool is to serve as a representation of the UWwS being designed. The tool should allow the user to identify cause-effect relationships between competing designs scenarios and the selected PIs of the UWwS. As stated in Chapter 2, a number of integrated modelling tools exist; however, they appear not to be suitable for this research. One of the main limitations is the fact that the freeware licensed tools are normally strong in one component and oversimplified in another UWwS component. They miss what Raunch *et al* (1998a) call a balance in the complexity of the models that represent each component of the integrated model. In addition, commercially available integrated software often lacks the connexion with optimization algorithms. Since their source codes are protected; hence, limited modifications can be done to customize those tools for the requirements of this research.

Considering above limitations of existing software, we decided to develop a customized integrated UWwS modelling tool. The integration is achieved by linking state of the art, freeware software for each component (sewer, WwTP, river) of the UWwS. The requirements of the model, the selected software, and the description of the integration of them are described in following sections.

3.4.1   Requirements of the integrated urban wastewater modelling tool

In general, the modelling software should represent the impact of the decision variables (design parameters and control parameters) on the performance of the urban wastewater system measured with the selected indicators. In other words, the model should allow the designer to modify the design parameters and should generate the data required to calculate indicators related with water quantity, water quality and cost. The criteria considered in the selection of the modelling software are listed below:

- In order to use the accumulated knowledge of modelling software for UWwS components, existing and validated software were selected to be linked into the integrated modelling platform.
- Considering that the modelling tool is to be used in UWwSs for which there is not many data available, a process-based model was preferred instead of data driven modelling approaches. In addition, deterministic modelling approaches were chosen instead of stochastic approaches. The selection was based on the fact that data driven and stochastic approaches would require a larger set of data for model development.
- As MoDeCo approach requires the analysis of the dynamic interaction between UWwS components. Thus, neither static nor a steady-state modelling approach is sufficient. Therefore, dynamic modelling tools were selected to be able to represent transient processes (e.g. backwater effects in sewers).
- Preferably, the model should allow the simulation of single events and time series for long term simulations. The integrated approach may require the simulation of longer periods to assess impacts in the receiving system. Long term continues simulation may also be needed to assess accumulative pollution impacts in the receiving system, or if statistical indicators are used to assess impacts in the receiving system as indicated by the

UPM Manual (FWR 1994). This research addressed only acute pollution impacts in the receiving system. Therefore, the single event or a combination of events stitched together was used. Perhaps long continues simulations may be used in the post-processing of the optimum design solutions.

- When selecting the software, care has to be taken to match the complexity of the modelling tools for each component of the UWwS (Rauch, *et al.* 1998a). The modelling tools should allow the user to select different levels of complexity depending on the problem being addressed.
- The integrated tool should allow the designer to simulate control strategies as part of the integrated design of the system. For this research local control strategies were preferred. A structure is operated using local control when the data used to decide the control action is gathered in the vicinity of that structure; it does not depend on the communication with the other facilities or other parts of the water system (Lobbrecht, *et al.* 2010). In terms of modelling requirements this facilitate the connexion of the models of each sub-component because there is no need for communication of data in run time. Therefore, the simulation can be done in sequential mode. That is, all time steps of the simulation of one component must finish before the next component can be modelled. If integrated control strategies are preferred then the model of each component must be done in parallel (Schütze, *et al.* 1999).
- Due to the intended application of optimization algorithms for the determination of the design and control parameters, the modelling software should be easily embeddable into the optimization procedure, without losing the option to be run as stand-alone modelling software if needed.
- Cost and availability are additional criteria. Considering the interest to apply this research in situations with limited resources, freeware software is preferred over the commercial versions.

Considering the criteria presented above with the software available when this research started, the following modelling tools are selected for the integration:
- To model the hydrological process in the urban catchment and the transport in the sewer network, we selected the Storm Water Management Model (SWMM),
- To simulate the processes in the treatment plant two alternatives were implemented. The first alternative is the Activated Sludge Model ASM1 implemented by the author in MatLab, and the second is by connecting the STOAT dynamic modelling software for WwTPs.
- The processes in the river are modelled using the Water Quality Analysis Simulation Program (WASP).

A description of the selected software is presented in the following section.

3.4.2 Modelling sewer system

The Storm Water Management Model (SWMM) was developed by the United States Environmental Protection Agency (US-EPA). This modelling software is freeware licensed and can be downloaded from www.epa.gov website. Detailed information of the conceptual model and the user interface can be found in the user manual of version 5 (Rossman 2008).

SWMM is a dynamic rainfall-runoff simulation model used for single event or long-term (continuous) simulation of runoff quantity and quality from primarily urban areas. The runoff component of SWMM operates on a collection of sub-catchment areas that receive precipitation and generate runoff and pollutant loads. The main hydrological processes are represented in SWMM: rainfall interception, infiltration, percolation into groundwater, snow accumulation and evaporation. The model uses a nonlinear reservoir approach for routing the overland flow. The pollutants load is generated using the build-up and wash-off approach. Different particle sizes can be represented in the model and these are associated with the common water quality parameters (i.e. chemical oxygen demand, suspended solids, etc).

The routing component of SWMM simulates the transport of the runoff through a system of pipes, channels, storage/treatment devices, pumps, and regulators. SWMM conserve the mass balance of the water quantity and quality constituents of the catchment runoff and any external flows and water quality inputs from groundwater interflow, rainfall-dependent infiltration/inflow, dry weather sanitary flow, and user-defined inflows. SWMM also simulates the reduction in the concentration of water quality constituent through treatment in storage units or by natural processes in pipes and channels. The user can specify two different flow routing methods: kinematic wave or full dynamic wave. With the dynamic method the model can simulate various flow regimes, such as backwater, surcharging, reverse flow, and surface ponding.

In terms of control strategies the user can define dynamic control rules to simulate the operation of pumps, orifice openings, and weir crest levels. If-then-rules and PID controllers (proportional-integral-derivative) can be implemented. The sensors can be any attribute (e.g. flow, depth, head, etc) of nodes, conduits, storage, pumps, weirs or orifices in the network. The controllers act on any setting of the actuators (e.g. pump on/off, weir crest level, orifice opening percentage, etc).

In terms of PIs, apart from the usual hydrological and hydraulic variables (runoff volume, peak flow, water levels, velocities etc.), SWMM is able to compute indicators suitable for design and control. For instance, flow classification, surcharging in the sewer network, surface flooding and pumping volume and power usages. The user-defined water quality variables can be also specified as an output of the model.

The availability of SWMM source code increases the potential for interaction with other software tools. Several interfaces have been developed which use freeware SWMM as mathematical engine. In addition, the formatted text input file facilitates the modification of the design and operational parameters in the sewer network. Even though, the majority of the outputs are in a binary files it is possible to generate most of the outputs as a formatted text file. Thus, the output text file is used to compute indicators or to generate time series of input data for the subsequent model of the UWwS (i.e. the treatment plant model and the river model). In other words, having the inputs and outputs in a text file facilitates the interaction with other software tools selected for the integrated modelling of the UWwS.

3.4.3 Modelling the wastewater treatment plant

At the beginning of the research none of the available models of the WwTP fulfilled the requirements because most of them were commercial software and modifications in the code are not allowed. Therefore, linking the other components of the UWwSs and the optimization algorithm was not possible. The decision was made to develop a code in MatLab to simulate the treatment processes. The model describes the activated sludge and the clarification processes.

The activated sludge processes are represented using the Activated Sludge Model No1 developed by the IWA task group on mathematical modelling for design and operation of biological wastewater treatment (Henze, et al. 2000). Five combined steered tank reactors (CSTR) in series are included for that purpose. The anoxic zone is represented by two CSTR and the aerated reactor is represented by three CSTR. The secondary settler is modelled using the Tackas model without biological reactions. The settler is subdivided in 10 layers and the feed point is located in the sixth layer. Two return flows are included: the nitrate internal recycles from the 5th to the 1st reactor and the return of activated sludge from the underflow of the secondary settler to the front end of the plant. The excess of activated sludge is wasted continuously from the secondary settler underflow at a rate.

The model was developed following the examples presented by Olsson and Newell (1999) for modelling WwTPs. The code includes simplified capabilities for control of oxygen dissolved in the aerated reactors using the air flow rate. The WwTP code was validated using the data and results from the Cost Project. The model reproduces the result of the benchmark model developed for Cost project (Copp 2002). As the model was developed in MatLab, reading text files inputs from SWMM was relatively simple and the data transfer to the river model using text files as an output was a direct solutions. Nevertheless, the significant advantages of having code the model is pay back by the fact that more complex schemes of the WwTP could not be incorporated. For instance if the interest was in modelling the Phorsphorous removal the model need a significant upgrade to include the descriptions of this processes (i.e. ASM2d). Considering this situation, we seek for alternative models that still being available as freeware software, allow the selection of different type of models of the WwTP. In 2010 STOAT software becomes freeware software and we adopted as on of the components of the integrated modelling tool.

The STOAT - dynamic modelling software for WwTP is developed by the Water Research Centre (WRc plc) from UK. STOAT has been developed as part of the UK Water Industry's Urban Pollution Management programme. During the UPM programme STOAT was extensively validated against sewage works' data. Recently this modelling software was released as a freeware licensed and can be downloaded from http://www.wrcplc.co.uk website. Detail information of the conceptual model and the user interface can be found in the set of manuals provided with the software: Installation and User Guide, Process Model Descriptions and Tutorials Guide (WRc plc 2010).

STOAT is a PC based computer modelling tool designed to dynamically simulate the performance of a wastewater treatment plant. The software can be used to simulate individual treatment processes or the whole treatment works, including sludge treatment processes, septic tank imports and recycles. The model enables the user to optimise the response of the works to changes in the influent loads, treatment capacity or process operating conditions (WRc plc 2010). STOAT contains a wide range of features, including: models all common treatment processes, offers both Biochemical Oxygen Demand (BOD) and Chemical Oxygen Demand (COD) models, support for user-written models, and allows simplified sewer modelling and easy data transfer to other modelling tools.

In terms of control strategies the user can define dynamic control rules to simulate the operation of pumps, flow dividers, air blowers and chemical dosing. If-then-rules and PID controllers can be implemented. The sensors can be defined as any state variable (e.g. COD) of the processes being modelled. The controllers act on any setting of the actuators (e.g. pump on/off, air blowers and divider flow rates, etc). In terms of PIs the model can estimate basic statistics (e.g. 95 percentiles); however, the output generated by the model can be easily manipulated to estimate for instance performance efficiencies or operational costs.

Since STOAT was developed within the framework of the UPM programme, it has build in capabilities to model the sewer system using a simplified approach (SIMPOL) and recently a model of the receiving system has been include to mainly assess pollution impacts in rivers (River model No1). It has been also connected with more complex models like MOSQUITO for the sewer and MIKE 11 for the receiving system. The exchanging data with models for rivers and sewers has been improved by the development of the OMI file function that is used to create a configuration file for use within OpenMI simulations.

Data project input is available through the user graphical interface. The project data are stored in Microsoft Data Base (MDB) files. Some of the time series of data can be input as a text file. The outputs of the model are also stored mainly in MDB files. However, some variables can be selected to be printed in text files. In this research, to modify the input parameters of the STOAT model we modify the MDB files. In addition, the text files were used to input time series from the sewer model and to generate time series as text files for the river model.

3.4.4 Modelling rivers

The Water Quality Analysis Simulation Program (WASP7) is developed by the United States Environmental Protection Agency (US-EPA). This modelling software is freeware licensed and can be downloaded from www.epa.gov website. Detail information of the conceptual model and the user interface can be found in the user manual (Wool, *et al.* 2005).

WASP is a dynamic compartment-modelling programme for water systems, including both the water column and the underlying benthos. WASP simulates the flow routing processes in a receiving system represented in 1, 2, or 3 dimensions. In a river represented by a one-dimensional, branching network, WASP is capable of internally calculating flow routing using the kinematic wave formulation. The model also can solve the routing equations using the fully dynamic formulation; therefore is possible to represent the flow in ponded segments, weir overflows and bi-directional flows caused by backwater effects.

WASP simulates the transport and transformation of a variety of water quality components. Water quality processes are represented in special kinetic subroutines that are either chosen from a library or written by the user. WASP is structured to permit easy substitution of kinetic subroutines into the overall package to form problem-specific models. WASP comes with two kinetic sub-models to simulate two of the major classes of water quality problems: conventional pollution (involving DO, BOD, nutrients and eutrophication) and toxic pollution (involving organic chemicals, metals, and sediment). In terms of PIs, WASP can generate time series of data of the state variables being modelled. Thus the calculation of the indicators (e.g. maximum COD or minimum DO) can be computed from the results of the model.

WASP can be linked to hydrodynamic models. Upon execution of a WASP input dataset using these option the hydrodynamic linkage file must already be created and exist in the directory that the input dataset resides. The file must have the extension of *.HYD. The hydrodynamic linkage dialog box allows the user to select a hydrodynamic linkage file. The hydrodynamic linkage file provides flows, volumes, depths, and velocities to the WASP model during execution. There are several hydrodynamic models that have been linked with WASP. The models include: DYNHYD5, RIVMOD, EFDC and SWMM (Wool, *et al.* 2005).

WASP stores the model-input data in individual files, with extension WIF, WASP input file. The input file is binary which allows for rapid saving/retrieving of information from the user interface. In general, the schematization and the instantiation of the model can only be done through the WIF file. That is through the user interface of WASP. However, the time series of data for the boundaries and the discharge points (i.e. WwTP discharge) can be imported from excel files. This feature allows us to link the effluent of the WwTP and the discharge point of the sewer system with the model of the river.

In terms of output files, WASP generates also binary model output files. Those files can be read by a Post-Processor. The Post-Processor reads the output files created by the models and displays the results in two graphical formats. In addition, the user may export the data that is used to generate the active x/y plot to an external file that can be in the format of a comma delimited ASCII files. This output text files are used in the integrated model to read the time series of the estate variables for the calculation of the pollution impacts in the receiving system (river).

3.4.5 Integration of the modelling tools

Since the approach used to link the models is sequential, the main interaction between the modelling software is the input and output files. To generate the input and output files in the adequate format for each component a number of interface algorithms were written. Those interface algorithms were developed buy the author using Delphi and MatLab. Figure 3.3 shows the scheme of the integrated modelling tool and the different interface algorithms required to integrate the models of the UWwS components.

Due to the different state variables in the models selected, the interface algorithms also include the transformation equations to transfer the state variables from the sewer to the WwTP and from the WwTP to the River. The composite state variables are calculated using the equations proposed by Copp (2002). In addition to the interface algorithms, there is the need to create the PIs. Most of the results of the models are raw data that require some mathematical calculations to be transformed in to the indicators that represent the objectives being optimized.

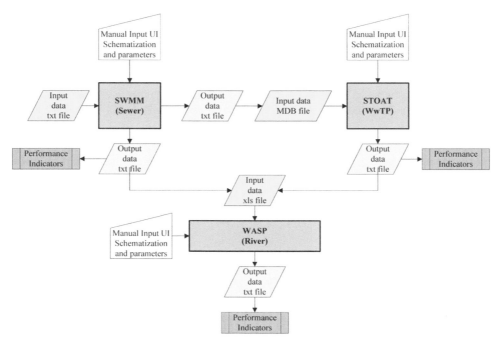

Figure 3.3 Scheme of the integrated UWwS modelling tool

3.5 Multi-Objective Optimization Tool

3.5.1 Selection of the algorithm for optimization

The literature reviewed in chapter 2 show that one of the best alternatives to solve Multi-Objective Optimization Problems (MOPs) in UWwSs is the use of Multi-Objective Evolutionary Algorithms (MOEAs). A number of MOEAs have been evolving since the first algorithms were developed in the mid 80's. Some of the most representatives MOEAs are: Pareto Archived Evolution Strategy (PAES) developed by Knowles and Corne (1999), the Pareto Envelope-based Selection Algorithm (PESAII) developed by Corne, *et al.* (2001) The Strength Pareto Evolutionary Algorithm (SPEA2) developed by Zitzler, *et al.* (2001) and the Non-dominated Sorting Genetic Algorithm (NSGA-II) developed by Deb, *et al.* (2002). A detailed analysis of these and other MOEAs can be found in Coello Coello, *et al.* (2002).

The main difference between PAES, PESA, SPEA and NSGA is the exploration of different ways to do selection and population maintenance in multi-objective spaces. The SPEA2 and NSGA-II are two of the most prominent MOEAs used when comparing the latest versions of MOEAs available. Prevalent in these two MOEAs is the fact that they preserve non-dominated solutions in the Pareto (i.e. elitism) and they rely heavily on their density estimator mechanisms. The density estimator is based on the volume of the hyper-rectangle defined by the nearest neighbours. The estimators (e.g. crowding distance) are mechanisms used to measure the diversity in the population of possible solutions.

There is no "best" MOEAs, as it has been demonstrated in the No Free Lunch Theorems (NFL) by Wolpert and Macready (1997). Certain MOEAs have been experimentally shown to be more effective than others for specific MOP benchmarks and certain classes of real-world problems. NSGA-II has proven to be a very powerful technique to find optimal solutions for many optimizations of UWwS. The potential of NSGA-II has been demonstrated in optimizations of drainage systems (Barreto, et al. 2006), design of UWwS considering water quality objectives (Muschalla, et al. 2006) and integrated control strategies of UWwSs (Fu, et al. 2008). Therefore, in this research we select the NSGA-II algorithm as the main optimization tool. Since the focus of this research is not on the optimization algorithm itself, we decide to use NSGA-II and analyze the implications when applied to the optimization of UWwSs. However, the approach proposed here may works with any other MOEA. The following section describes the NSGA-II and the steps required to implement it in the general MoDeCo framework.

3.5.2 Description of non-sorted genetic algorithm NSGA-II

NSGA-II is an improved version of the non-dominated sorting algorithm (NSGA) proposed by Srinivas and Deb in 1994. The improvements in the code are: i) the complexity of the algorithm was reduced from $O(mN^3)$ to $O(mN^2)$ where m is the number of objectives and N is the size of the population; ii) an elitist selection method was include to speed up the identification of Pareto solutions iii) the sharing parameter used to maintain diversity in the solutions was eliminated by including a crowding distance comparator that does not require the definition of additional parameters (Deb, et al. 2002).

NSGA-II is an implicit building block MOEA. That is, NSGA-II builds a population of competing individuals (solutions), ranks and sorts each individual according to non-domination level. An individual p1, within a population, dominates another individual p2, if and only if it performs as well as p2 with regard to all objectives and strictly better in at least one objective. The non-domination sorting approach divides the population into different ranks. The individuals who do not dominate each other but dominate all the others in the population are assigned rank 1, the best individuals in the population. Amongst the remaining individuals in the population, the individuals who do not dominate each other but dominate all the others are assigned rank 2. The same procedure is repeated until all the individuals are assigned a rank (i.e. blocks are built). Then, NSGA-II uses evolutionary operations (selection, crossover and mutation) to create new solutions, named child population. The parents and child populations are combined to form a combined population of possible solutions.

NSGA-II then conducts niching by adding a crowding distance to each member. The crowding distance of each individual is calculated by the average Euclidean distance between the individual and those adjacent individuals in the objectives space. It uses this crowding distance in its selection operator to keep diversity in the population by making sure each member stays a crowding distance apart. This also helps the algorithm to explore better the objective function space. Thus, the selection process at various stages of this algorithm is based on two criteria: non-domination rank and crowding distance. First is the rank, thus, an individual with lower rank is selected. Otherwise, the individual with greater crowding distance is preferred if both individuals belong to the same rank. In each generation, new individuals are generated through crossover and mutation operations on selected parents. The pseudo code of the NSGA-II is shown bellow:

1. For t = 0
2. Create randomly a parent population P_0 of size N
3. Estimate the respective function values $F_k(p)$ for each $p \in P_0$.
4. Sort P_0 using the non-domination sorting approach
5. Rank the solutions by assigning a fitness value equal to the non-domination level (fitness rank = 1 is best assuming minimization).
6. Create a child population Q_0 of size N using selection, crossover, and mutation functions.
7. Assign the respective functions values $F_k(r)$, where $r \in R_0$ and rank the solutions using the non-domination level
8. For t ≥ 1
9. Combine parent and children population $R_t = P_t \cup Q_t$. R_t is size 2N.
10. Sort R_t according to the non-domination approach on $F_k(r)$.
11. Calculate crowding distance for $F_k(r)$ where $r \in R_t$.
12. Sort in descending order using crowding distance comparator criteria
13. Form a new parent population P_{t+1} by selecting from R_t first considering the non-dominated solutions and second the crowding distance comparator criteria
14. Create a new child population Q_{t+1} of size N using binary tournament selection, crossover, and mutation operators.
15. Loop till ending criteria is achieved (e.g number of generations).

For this research we use the implementation of the NSGA-II available in the MATLAB toolbox for Genetic Algorithms. Modifications of the code are explained in the following chapters as required by the specific optimization problem being solved.

3.5.3 Steps in the optimization

The following steps were used to implement the optimization process in each case study. The steps are illustrated with examples.

a) Define the system boundaries and the interfaces with the outside world
- Define the system boundaries. Boundaries may be defined at the urban catchment by selecting the sub catchments of interest. In addition the interactions with other components like the WwTP and the receiving system must be defined.
- Define the inputs: manipulated variables (also called degrees of freedom) and disturbances (which we cannot manipulate). The manipulated variables in a drainage design are normally associated with the pipe sizes, slope, cover depth, storage volume, etc. One of the main disturbances of a drainage system is the precipitation.
- Define time scale of interest. The time is highly dependent on the system being designed. In a drainage design, the time of concentration is important for the selection of the precipitation event used to assess the alternative solutions. This time scale may be in the order of hours. However, if the interest is also in reducing pollution impacts in the receiving system the time scale may be increased to the order of days or even weeks.
- Define the system state variables, which in turn determine the output variables to be optimized. The state variables in a drainage design are associated with water quantity (e.g. flows, flooding volumes) and/or water quality parameters (e.g. organic mater, suspended solids).

b) Define the decision variables (degrees of freedom)
- They are the design variables or the control set points.
- They must have an impact in the objectives.

- The degrees of freedom (n) define the n-dimensional space of the decision variables.

c) Define the objective functions
  - Objective function has to be able to reflect the consequences of different changes in the decision variables.
  - Each objective must have an indicator that can be measured. The indicator can be the state variables of the modelling software (e.g. flooding volume) or can be estimated as a function of the decision variables (e.g. cost of the pipe network).
  - In this research we use three type of performance indicators:
    - Water quantity objectives. These types of indicators are normally a result of the modelling tools or are estimated using the results of the model. Examples are flooding volumes, combined sewer overflows, or influent flows to the treatment plant.
    - Water quality objectives. These indicators are also based on the results of modelling tools. Normally they require an additional calculation. For instance, to estimate the minimum or maximum over the standard value. It is also possible to consider a combination of the parameter's critical value and the period when the standard is exceeded. An example of composite indicator can be found in Schütze, et al. (2002).
    - Cost objectives. In general terms there are two types of costs indicators, those related with the initial investment (capital cost) and those related with operational cost. Normally the investment cost is not the result of the modelling but estimation based on the decision variables. The operational cost is related with the use of resources. This cost may use modelling results in combination with mathematical expressions in function of the decision variables. Typical indicators of operational cost are energy (for pumping water and recycles and compressing air), chemical additives and manpower. In the case of optimum design the cost indicator should reflect a balance between capital and operating cost.
  - A metric to judge the sustainability of different options will facilitate the post-processing of the optimized solutions and the final decision making with the stakeholders.
d) Define the constraints
  - Define a feasible region for the process and the optimization. There are three types of constraints:
    - External constraints: effluent discharge limits and other environmental standards, safety, political policies.
    - Process constraints: mass and energy balances, flow rates and concentrations must be positive.
    - Equipment constraints: volumes, maximum and minimum flowrates, some equipment can only be on-off.
  - Process and equipment constraints may be mathematically defined, so we optimize a mathematical model and then apply the results to the system. This approach is called off-line optimization.
  - If the process is optimized on the run, the approach is called on-line. The constraints are defined by the process itself.

e) Response surfaces
  - It is wise to perform some response surface mapping before attempting an optimization.
  - Response surface will enable assessment of the sensitivity of the objectives function to the degrees of freedom, and to detect correlation between degrees of freedom

- Three dimensional contour or surface plots of the objective function in the feasible region should be examined. These can be plotted against degrees of freedom two at a time using a fairly coarse grid of operating points.
  - A very flat surface relative to a particular axis indicates a lack of sensitivity and the corresponding degree of freedom is best removed.
  - Values at an angle indicate correlation between degrees of freedom. It maybe possible to derive a mechanistic or empirical relation between the parameters and therefore remove one as a degree of freedom
  - Care has to be taken so that this relationship does not change with time. Otherwise, it has to be updated when system or objective function parameters change.

f) Setting up the optimization algorithm
  - The user of the algorithm must decide which functions are included (i.e. selection, mutation and crossover), and must specify the parameters that control the MOEA's search: population size, number of generations, probability of mating, probability of mutation and stopping criteria.

g) Selection of initial population
  - One of the settings that speeds-up the optimization process is the use of known sub-optimal solutions in the initial population. The best known solution corresponds to the pre-design of the system; therefore it can be use as a seed in the initial population.

3.6   Case Studies

Two case studies were selected one in Colombia and one in Netherlands. The case studies are used to apply the methodology proposed in this chapter and answer the second research question: identify the main benefits and drawbacks of MoDeCo approach? In both cases, a comparison is carried out between the design based on traditional approaches and the optimum design found using MoDeCo.

Case study of Cali – Colombia:
Cali has a population of 2.07 millions of inhabitants (year 2005), and the city is growing at a significant pace. The local government designated an area for the future expansion of the city. The case study is developed in a sub-catchment of 70 ha of the expansion zone. This specific area is to be developed within 25 years. Therefore, this area requires the design of the urban wastewater system. The design problem of this case study is the most complex because degrees of freedom are huge. The case represents a challenge for MoDeCo approach. The objective is to identify the optimum design solution for the sewer network considering the interaction with the treatment plant and the river. In addition, the design includes the optimization of the operational strategy. This case study is presented in Chapter 4.

Case study of Gouda – The Netherlands:
Gouda has a fully developed UWwS including combined and separated sewer network, an activated sludge WwTP and the system discharge to the canals of the city and in the Hollandse Ijssel River. Considering that the system is in place, the objective of this case study focus in the optimum operation of the system. This is realized by the optimization of the functional design of the treatment plant. The challenge is to use the information generated in the sewer network to better control the treatment plant. This case study is presented in Chapter 5.

3.7   Conclusions

The development of the integrated UWwS modelling tool demanded most of time and effort in the development of this approach. Even though, the integrated modelling tool matched the requirements of this research and the needs of case studies, there are still some limitations:

- There is no interface, thus, user must use MatLab as the general command window. In addition, the user must set-up each model for each component, and input model parameters and initial data through the user interfaces of SWMM, STOAT and WASP.
- The demand for data to feed into the integrated modelling tool appears to be one of the major limitations of the approach. However, if the objective is the design of the components (Sewer or WwTP) there is no need for detailed calibration, except for the receiving water system (River).
- The sequential combination of software for each sub-component of the UWwS brings consequently the following limitations: different state variables, limited flux of information between models of subsystems and long calculation times.
- The selection of the model complexity appears to depend on the problem being addressed. As a basic rule a balance of complexity in each sub-component should be follow. In addition, the author suggests applying the Einstein's razor which paraphrased for this research could be: "the integrated model should be as simple as possible but not simpler, as loss the accuracy of the representation of the UWwS processes".
- The MoDeCo approach itself need to be tested in many different cases. Therefore, this is just the beginning, a small contribution towards a better design and operation of UWwS.

Nonetheless, MoDeCo has a great potential for the optimum design and control of UWwS considering the interaction between components. In the following chapters this potential and the main benefits and drawbacks will be explored with the implementation of the case studies.

4 Design of an Urban Wastewater System for Cali

4.1 Introduction

One of the most common ways to include models in the design of UWwSs is in the dynamic assessment of different scenarios based on a steady state design. Since there is a large range of possible slopes, diameters and roughness coefficients of pipes, only a small number of combinations of these parameters are usually analyzed in traditional design processes. Thus, the design reduces to the evaluation of a few scenarios that depends on the expertise and skills of the designer. In addition, the study of various design options for the sewer system often ends at the overflow structures and the wastewater treatment plant (WwTP) inlet, without taking into account the role and function of the treatment system and the receiving water body (Butler and Schütze 2005). The transfer across the interfaces of each subcomponent is characterized by static rules. For example, the flow to the WwTP under wet-weather conditions is limited to a value the order of twice the peak dry-weather flow (Rauch, et al. 2005). The operation of UWwSs is often based on empirical and static rules that do not take into account the complexity and the dynamics of the influent water quality and quantity. The disturbing feature of this sub-optimum operation is that parts of the system might have considerable spare capacity while, at the same time, other parts are overloaded. This is where real time control becomes an option. Manipulate the system such that its capacity could be used better in order to achieve improved performance of the system.

In the past three decades, efforts to develop optimization models for the optimal hydraulic design of a branched sewer system have been extensive. Multiple examples of the optimal hydraulic design are presented by Weng and Liaw (2004). However, the optimal hydraulic design based on a predetermined layout does not guarantee the optimal design of the system. The need for the optimal performance of the system considering the reduction of pollution impacts leads to few examples. This implies the need for a more integrated view of the system as is illustrated in the case study developed by Muschalla et al., (2006) which optimizes the design of an urban drainage system using water quality objectives and an evolutionary algorithm to determine the optimum settings.

Many more applications can be found in the optimization of control strategies for an integrated UWwS based on water quality criteria. Rauch and Harremoës (1999) present a novel approach to control the whole system: sewer system, treatment plant and receiving water with the aim of achieving the minimum effects of pollution. The application of nonlinear model predictive control by means of a genetic algorithm reveals excellent results for hypothetical problem sets. Schütze et al (2002) combines an integrated simulation model for the sewer system, WwTP and receiving water body with several optimization methods, among then a Multi-Objective Evolutionary Algorithm which they conclude is the most appropriate for the problem. Vanrolleghem et al. (2005), describes an approach to optimize directly the river water quality based in a control strategy which exploits the interactions between the different subsystems. Fu et al. (2008) presents one of the latest examples of using water quality objectives in the river as criteria for optimal control of the maximum outflow rate of a storage tank, the maximum inflow to the treatment plant, the threshold triggering the emptying of storm tank and the return of activated sludge in the treatment plant.

The approach proposed Model Based Design and Control (MoDeCo) in this thesis extends further the optimization technique by combining the optimization of the hydraulic design of the UWwS components with the control rules that define the functional design of the system. To identify the best performance with a minimum cost, MoDeCo solves the problem using a multi objective optimization approach. The design problem of an urban drainage system for Sector 1A of the expansion zone of Cali (Colombia) was implemented as a case study for the research.

4.2   Description of the Urban Wastewater System of Sector 1A in Cali

4.2.1   Description of the urban catchment and drainage

The case study is situated in the southern part of Cali city in Colombia. The area was defined by the municipality as the expansion zone for the city (Figure 4.1). The expansion zone covers an area of 1652 ha and it is expected that it will accommodate approximately half a million inhabitants by 2030. Within the expansion zone, we selected the Sector 1A to be used as a case study. The sector 1A covers an area of 70 ha and will provide housing for approximately 21694 inhabitants.

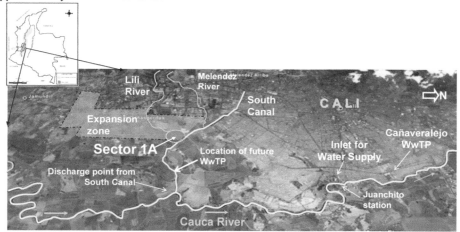

Figure 4.1 Schematic overview of expansion zone of Cali and location of Sector 1A.

(Map source: Google Earth 2009)

The utility company Empresas Municipales de Cali EICE ESP (EMCALI), developed the feasibility study for urban wastewater management in the expansion zone (EMCALI and Hidro-Occidente SA 2006). The population density, water consumption and wastewater production are defined in the feasibility study and presented in Table 4.1. The wastewater production per capita is estimated as 80% of the water consumption. The layout of the roads and the sub-catchments are shown in Figure 4.2. A more detail description of the sub-catchments is included in Table 8.2 (Appendix 8.2).

Table 4.1 Characteristics of the sector 1A, expansion zone of Cali.

Characteristic	Value	Units
Area	70	ha
Sub-cathments	32	
Population density	310	h/ha
Water consuption	220	l/h/d
Wastewater production	176	l/h/d

4.2.2   Description of the wastewater treatment plant

The treatment of the wastewater produced in the expansion zone is a critical issue, because the final receiving system is the Cauca River, which is at the same time the main source for the water supply system of Cali. In the feasibility study different options were analyzed for the treatment and disposal of the wastewater (EMCALI and Hidro-Occidente SA 2006). According to the analysis elaborated by EMCALI and Hidro-Occidente SA, the best two alternatives in order of eligibility are: 1) pump the wastewater towards the north part of the city and treat it in the existing WwTP Cañaveralejo, and 2) build a new wastewater treatment plant to the south of the city based on the activated sludge process with removal of the nitrogen and the disinfection of the effluent before discharging it to the South Canal (Figure 4.1).

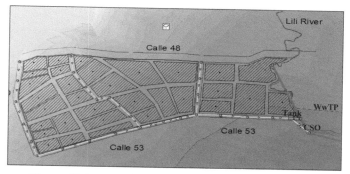

Figure 4.2 Layout of the sector 1A, expansion zone of Cali.

According to the analysis of alternatives presented by EMCALI and Hidro-Occidente (2006), it seems that a significant percentage of the wastewater produced could be pumped and treated in the Cañaveralejo WwTP. However, it also seems that some parts of the expansion zone are less suitable to follow this alternative because they are located in the lowest terrain, implying more pumping. This seems to be the case with Sector 1A. Therefore, for the case study we decided to consider a new wastewater treatment plant in the south. The hypothetical treatment plant is projected only for the wastewater production of Sector 1A, and the location is defined at the outlet of the urban catchment, discharging the effluent to the Lili River, as shown in Figure 4.2. The treatment is based on the activated sludge process and includes the biological removal of nitrogen. The general overview of the components is presented in Figure 4.3.

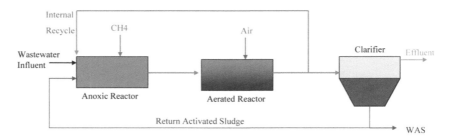

Figure 4.3 General scheme of the wastewater treatment plant

4.2.3 Description of the Lili River

According to the feasibility study for the drainage system, the sector 1A can be drained to the River Lili by gravity. The main channel of the river is approximately 20 km long from its source in the upper part of the Andean chain to the mouth in the South Canal in the valley of the Cauca River (Figure 4.4). A profile of the river is presented in Figure 4.5. According to the catchment elevation, it has distinctive characteristics of the land use: from a more pristine upper part to a heavily populated area in the valley.

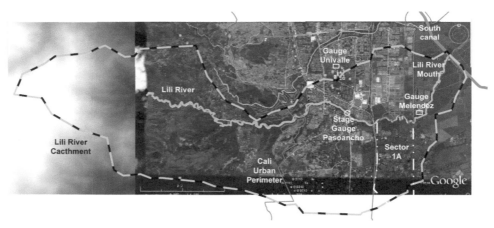

Figure 4.4 River Lili catchment

The catchment can be subdivided into three zones according to their characteristics:
- Upper catchment: is a hilly area with steep slopes in the order of 50%. The use of the soil is dominated by subtropical rainforest vegetation.
- Middle catchment: is the part of the catchment between the townships La Buitrera to the beginning of the Cali city's limits. The area is characterized for slopes the order of 15% and the predominant use of the soil for coal mine. Acid effluents from the mines are discharged into the river.
- Lower catchment: is the part of the river from the upper border of the city to its mouth in the south canal. The area is characterized by gentle slopes, and the soil is predominantly used for urban development. In this section the river crosses the city in an area that has experienced an increase in population density in the last decade. In this part the river receives storm and sanitary drainages from Cali (Discharges are shown in Figure 8.2,

Appendix 8). The last part of the river catchment crosses a farming area where the water is used for irrigation and as a receiving system of wastewater from sugar cane farms (González and Peñaranda 2004).

Sector 1A is located in the lower catchment. Therefore, the area of interest for the case study is the lower part of the catchment. However, the upstream flows and characteristics of the water influence the design and operation of the drainage of Sector 1A. In addition, the discharge of Lili River into the South Canal may be influenced by the water levels in the canal and in the Cauca River, which is the final recipient of the flow.

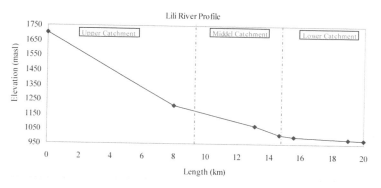

Figure 4.5 Profile of the River Lili

The main source of data for the river is the stage gauge located where the River Lili crosses the road Pasoancho, in Cali (Figure 4.4). The data collected at this station was used by the environmental authority Corporación Autónoma Regional del Valle del Cauca (CVC) to create the flow duration curve presented in Figure 4.6. The average daily flow that may occur 50% of the time is 0.38 m^3/s. The maximum annual peaks registered in Pasoancho gauge are presented in Figure 4.7. According to EMCALI and Hidro-Occidente (2006), the capacity of the Lili River is not enough to convey the current peak flow for any scenario that includes the effects of backwater in the South Canal. Due to the limited capacity of the main channel, the discharge of the drainage from the expansion zone will just make the situation worse. Therefore, the solution proposed by EMCALI and Hidro-Occidente is to increase the conveyance capacity of the last 5 km of the River Lili. This is a complicating factor for the drainage design, because the discharge to the River Lili may not be feasible by gravity alone in the case that peak flows coincide with high levels in the Cauca River and South Canal.

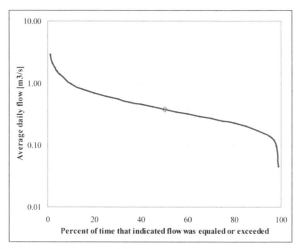

Figure 4.6 Flow duration curve for Lili River at gauge Pasoancho (Data: 1984 – 2008)

Source: (CVC 2009)

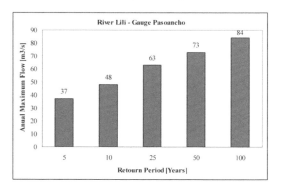

Figure 4.7 Annual peak flows for the River Lili at gauge Pasoancho.

Source: (DAGMA and UNIVALLE 2009)

In terms of water quality the information is rather scarce. However, some of the monitoring campaigns carried out by the local environmental authority Departamento Administrativo de Gestion del Medio Ambiente (DAGMA) show that the River Lili is negatively influenced by discharges from the city. The water quality objectives for the River Lili fixed by DAGMA are presented in Table 4.2. The results of two water quality monitoring campaigns are presented in Figure 4.8.

Table 4.2 Water quality objective for the River Lili

Parameter	Value	Unid
DO	4	mg/l
BOD_5	10	mg/l
TSS	15	mg/l

Source: (DAGMA and UNIVALLE 2009)

Three out of four water quality parameters analyzed in the River Lili show impairment of the water as it passes Cali. For instance, Dissolved Oxygen (DO) decreases below the 4 mg/l standard and the Biochemical Oxygen Demand (BOD$_5$) and Chemical Oxygen Demand (COD) tend to increase after the river crosses the city. The Total Suspended Solids (TSS) tends to decrease after the city, but this can be a consequence of sedimentation of the suspended material due to the flat slope of the main channel (Figure 4.5).

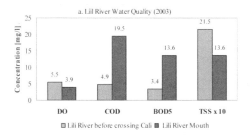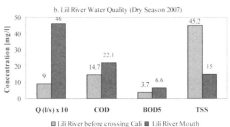

Figure 4.8 Water quality characteristics of Lili River

Source: (DAGMA and UNIVALLE 2009)

4.2.4 Formulation of the problem and objectives

Problem statement

The development of the expansion zone of Cali will increase the threat of the pollution in the River Lili and its final destination, the Cauca River. The receiving system is very sensitive because the Cauca River is the main source of water for the supply of 1.8 millions inhabitants. The assimilation capacity of the River Lili has been pushed to the limit; therefore there is no room for excess flows and pollution from the expansion zone of Cali. The challenge of this case study is to find a solution for the urban drainage and the treatment of the wastewater that limits the risk of flooding and at the same time reduces the pollution impacts.

Objectives

The main objective of the case study is to use MoDeCo approach to optimize the design and operation of the urban wastewater system for the expansion zone of Cali. The specific objectives are:

- Reduce the risk of flooding in the Sector 1A
- Reduce the pollution impacts on the receiving system caused by the urban development in the Sector 1A.

Aim of the case study

The aim of this case study is to demonstrate that with the use of MoDeCo approach it is possible to find an optimum design and operation of the UWwS in Sector 1A that will performs better that the design using traditional approaches.

4.3 Data and Methods

Summary of the research methodology

The general methodology is based on the model based design and control (MoDeCo) approach proposed in Chapter 3. The approach pushes further the optimization technique by combining the optimization of the hydraulic design of the UWwS components (sewer pipes, storage volumes, CSOs) and the control rules that define the functional design of the system. To identify the best performance with a minimum cost, MoDeCo solves this problem using a multi-objective optimization algorithm. First, the system is pre-designed and the operational strategy is predefined. A model is constructed based on the designed components and operational rules. The model run and the performance indicators are estimated for flooding events, impacts in the receiving system and cost. These three indicators form the objective functions that are optimized using the multi-objective optimization algorithm. The optimization routine assesses the performance of the system and creates new design parameters (diameters of pipes, slopes and operational strategy), thus generating a new scenario to be assessed via the process based model. The optimization loops until it converges to a Pareto set of optimum solutions based on the objectives proposed.

Data collection and availability
The information required for the design of the drainage system is based on the data available from the feasibility study (EMCALI and Hidro-Occidente SA 2006). Density of the population, wastewater production, layout of the roads and topographic maps are provided by EMCALI. Data of precipitation is available at two gauges: Univalle and Melendez (see Figure 4.4). The gauge Univalle is part of the national system of monitoring, and the information was made available by IDEAM. For the gauge in Univalle we have two years of data with a frequency of 10 minutes and the intensity-duration-frequency (IDF) curve. This is the main source of precipitation data used for the design of the sewer network. The gauge, Melendez, is part of the private monitoring network of the sugar cane farmers association. The daily average precipitation data for this station is freely available on the web.

Because this is an area of the city under development, there is no wastewater to be characterized by monitoring. However, data from Cañaveralejo WwTP was used to estimate the possible characteristics of the wastewater produced in the expansion zone. The data for the river is scarce and the main source is the report presented by DAGMA and UNIVALLE (2009). The research project lacked resources for a monitoring campaign, so the case study must be considered as hypothetical in which some simplifications and assumptions were made as explained below.

Data analysis and modelling tools
The data available was analyzed using MS Excel and MATLAB tools. Geographical and topographical information of the area was manipulated using AUTOCAD and ArcGIS.

The sewer system was modelled using the Storm Water Management Model (EPA SWMM 5.14) from the US, Environmental Protection Agency (Rossman 2008). The WwTP was modelled using MATLAB code developed by the author based on the Cost Simulation Benchmark Model (Copp 2002) as described in Chapter 3. To simulate the biochemical processes the Activated Sludge Model No 1 (ASM1) (Henze, *et al.* 1999) was selected. For the simulation of the river, the Water Quality Analysis Simulation Program (EPA WASP 7.3), from the US Environmental Protection Agency (Ambrose, *et al.* 1993) was used.

For the multi-objective optimization the Non-Sorted Genetic Algorithm (NSGAII) developed by (Deb, *et al.* 2002) and implemented in the MATLAB toolbox for Genetic Algorithms was used. For the integration of the models and the analysis of the results, pieces of code were developed in Delphi and MATLAB.

Assumptions

- We assume that the integrated model developed, even though it can not be calibrated, represents the behaviour of the UWwS in terms of trend and order of magnitude of the water quantity and water quality components.
- We assume also that there are no commercial or industrial zones in Sector 1A; all the area is considered for family housing and educational institutions.
- For the case study we assume that the sewer has a free discharge to the Lili River. In practice, the whole drainage system can be affected by backwater caused by the water level in the Cauca River. However, backwater effects in the sewer caused by the river are not considered. We assume that the conveyance of the river will be increased as proposed by the consultancy company in the feasibility study; and as a consequence, the river will not have problems receiving the flows from the UWwS.

4.4 Preliminary Design of the Urban Wastewater System

4.4.1 Design of the sewer network

Definition of sub-catchment characteristics and layout of sewer network
A combined sewer network was designed following the layout of the roads and the natural slope of the terrain. In order to characterize the sub-catchments, a digital terrain model (DTM) was created from contour maps of the project area using ArcGIS. For each sub-catchment the area, length, slope and drain direction were estimated. In addition, the DTM was used to define the ground surface elevations and the length of the pipes. The sub-catchments and the layout of the sewer network are presented in Figure 4.9. Notice that in the layout of the sewer network only pipes that follow the main roads have been schematized. The design of the sewer network is focused on these main pipes (the main trunk); secondary pipes are not included. In total the network includes 25 pipes and 26 manholes.

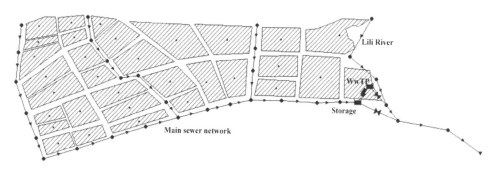

Figure 4.9 Sub-catchments and layout of sewer network for Sector 1A - Cali.

Estimation of dry weather flows
The dry weather flows (DWF) for Sector 1A correspond to the sanitary flow for the population expected by the year 2030. In Sector 1A there are no commercial or industrial areas that contribute to the DWF. The sanitary flows are calculated based on the information of the area per sub-catchment, the population density and the wastewater production. A peak domestic factor of 1.7 is included to calculate the peak sanitary flow used in the design. No

infiltration factor was included. The sanitary flows for each of the sub-catchments are presented in Table 8.2. The sanitary flow contributing to each of the manholes of the sewer network is presented in the hydraulic design (Table 8.3 Appendix 8.2).

Estimation of wet weather flows
For the preliminary design, the wet weather flow (WWF) was estimated based on the rational method (Q= C*i*A). Considering that the area is mainly residential a return period (TR) of 2 years was selected for the design storm. The Intensity - Frequency – Duration (IDF) Curve presented in Figure 4.10 was used to estimate the intensity (i) of storms for different durations (t). The run-off coefficient (C) was estimated assuming that 30% of the area is lawns. The composite C is presented in Table 4.3.

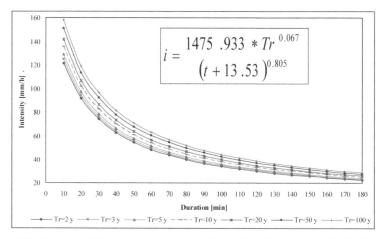

$$i = \frac{1475.933 * Tr^{0.067}}{(t + 13.53)^{0.805}}$$

Figure 4.10 Intensity - Frequency – Duration Curve from Gauge Univalle (1983 – 2003)

Source: (EMCALI and Hidro-Occidente SA 2006)

Table 4.3 Composite runoff coefficient C for Sector 1A – Cali

Catchment area		%	C
I	Residential	70%	0.65
II	Lawns flat	30%	0.16
	Composite C		0.50

Hydraulic design
The system was designed assuming circular pipes in concrete with a minimum diameter of 0.25 m. The minimum velocity for full pipe was defined as Vmin = 0.75 m/s and the maximum as Vmax = 5.0 m/s. The minimum cover depth was defined as 1 m. The inlet time was assumed to be 5 minutes. The estimation of the flow was based on the continuity equation (Flow = Velocity * Area). The velocity was estimated using the formula of Colebrook-White. The roughness coefficient of the pipes was fixed as k = 1.5 mm. The hydraulic design of the sewer network is presented in Table 8.3 Appendix 8.

Ancillary structures

At the downstream end of the sewer an on-line storage tank and a combined sewer overflow (CSO) were designed based on the Urban Pollution Management (UPM) Manual (FWR 1994). The setting of the CSO, that is the wastewater kept in the system to be treated, was based on formula A with additional storage (120 l/hab * Population) in order to limit the pollution impact on the receiving system. The design of the ancillary structures is presented in Table 8.4 of Appendix 8.2.

4.4.2 Design of the wastewater treatment plant

The WwTP was designed for the removal of organic matter and nitrogen and includes a primary settler, anoxic and aerated reactors and a secondary settler. The design is based on the methodology presented in Wastewater Engineering – Treatment and Reuse METCALF & EDDY (Tchobanoglous, *et al.* 2003). The treatment plant components follow the Modified Ludzack and Ettinger (MLE) configuration for BOD and Nitrogen removal. The system was designed for a peak flow of 2 times DWF. Wastewater characteristics used for the design are based on the average wastewater quality measured in 2006 at the inlet of the Cañaveralejo WwTP of Cali (Figure 4.11). The design of the wastewater components is presented in Table 8.5 of Appendix 8.2. Table 4.4 shows a summary of the characteristics of the designed components of the urban wastewater system.

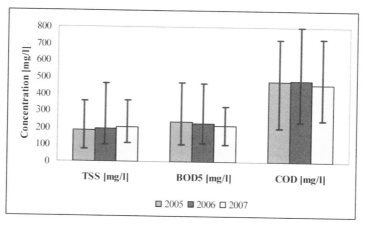

Figure 4.11 Influent characteristics of Cañaveralejo wastewater treatment plant

Table 4.4 Design characteristics of the sewer system and the wastewater treatment plant

Sewer			CSO + Tank			WwTP		
Characteristic	Value	Unit	Characteristic	Value	Unit	Characteristic	Value	Unit
Area	85	ha	Setting	420	l/s	Design capacity	49250	PE
Length	3.8	km	Storage	2604	m^3	Solid retention time	12.7	d
Population	25000	PE	Overflow	8480	l/s	Anoxic reactor	2000	m^3
Return Period	2	years				Aerated reactor	7125	m^3
Rainfall	91.5	mm/hr				Secondary settler	1728	m^3
DWF	75	l/s						
WWF	8900	l/s						

DWF: dry weather flow, WWF: wet weather flow and PE: population equivalent. Source: (Vélez, *et al.* 2008)

4.4.3   Functional design of the urban wastewater system

The aim of the operational strategy is to pump the most highly polluted wastewater (first foul flush) to the WwTP and to discharge through the CSO the less polluted wastewater. The main disturbances of the system are the precipitation events; and the actuators in the scheme proposed are the CSO and the pumping towards the WwTP. Therefore, the manipulated variable is the setting of the CSO, which is in our case, the flow pumped to the WwTP during the wet weather conditions.

The pumping station was designed with two pumps: one ideal pump that is continuously ON with a capacity equal to 1 DWF and a second pump with variable capacity to be turned-ON when a precipitation event occurs. The operational strategy may be implemented as If – Then rules, which include the water level in the tank that triggers a pump for the WWF and the capacity of the pump (flow setting to the WwTP). An example of the local strategy to control the pump is as follow:

$$\textit{If Node Storage Depth} >= 0.28$$
$$\textit{Then Pump2 Status = On}$$
$$\textit{And Pump2 Setting = 5.5}$$
$$\textit{Else Pump2 Status = Off}$$

4.5   Development of the Integrated Model

4.5.1   Sewer model

The implementation of the model for the sewer of Sector 1A - Cali is presented in this section. The main objective of the model is to understand the effect of the design variables on the performance of the system. Therefore urban catchments, pipes, manholes and ancillary structures are included in the model. The second objective is to optimize the functional design; which implies a description of the control of the process. Both the design and control require a modelling time step that allows the description of the processes in the urban catchment. The time step is associated with the retention time of the urban catchment; thus the time step is the order of minutes.

Schematization
The schematization of the model follows the structure of SWMM. The pipes are represented by objects called conduits and manholes are represented by nodes. SWMM also has objects to represent ancillary structures; thus, storage tanks, weirs for CSO, pumps and outflows are included and parameterized according to the design for Sector 1A.

Implementation of the sewer model in SWMM
The modelling of a sewer network can be divided in to two steps: the first is the transformation of rainfall into runoff and the second is the flow routing in the conduits. The rational method is a simplified model of the rainfall-runoff process. The run-off coefficient is a simple representation of the initial losses and continuous losses of precipitation volume. In SWMM these losses are represented by different processes. Initial losses for instance are represented by rainfall interception in depression storages and the continuous losses are represented by infiltration of rainfall into unsaturated soil layers. SWMM has three different models to approximate the infiltration process; for our case we selected the Horton method. This is perhaps one of the most used methods to represent the infiltration losses. The

parameters for the rainfall-runoff processes are the same for each of the sub-catchments and are presented in Table 4.5.

Table 4.5 Rainfall – Run-off parameters for SWMM model of Sector 1A

Parameter	Typical Value*	Value Selected	Unit	Comment
Horton parameters				Assumed Loam soils characteristics
Horton Max Infiltration rate	15 - 125	105	mm/h	Dry loan soil cover with vegetation
Horton Min Infiltration rate		3.5	mm/h	Saturated hydraulic conductivity of Loam
Saturated hydraulic conductivity	0.25 - 118		mm/h	Indication minimum infiltration rate
Decay constant for Horton	2 - 7	0.002	1/h	$0.00056 – 0.00139**\ s^{-1}$
Drying time	2 - 14	2	d	
Maximum infiltration volume		150	mm	
Catchment parameters				
Percent of impervious area		70	%	
Manning for impervious area	0.011 - 0.8	0.012		
Manning for pervious area	0.011 - 0.8	0.15		
Depth of impervious storage	1.3 - 2.5	2	mm	
Depth of pervious storage	2.5 - 7.6	3.5	mm	

* (Rossman 2008)
** (Huber and Dickinson 1988)

Once the model parameters are defined the next step is to define the input data for the precipitation. The rational method is a steady state representation of the system and the precipitation intensity used for the design corresponds to the one that produces the maximum flow from the catchment. SWMM model could well represent the steady state condition of the sewer. However, to exploit the capacity of the model to simulate the dynamics of the system, the inputs should be time series of precipitation or synthetic hyetographs. Due to the long calculations times of the sewer models, time series of precipitation will be prohibitively expensive for an optimization process. For the case study we use a combination of dry periods with synthetic hyetographs. There are different methods to define the synthetic hyetograph; the effects of three of them in the hydrograph produced by the model are presented in Figure 4.12.

The flow routing within a sewer pipe in SWMM is governed by the conservation of mass and momentum equations for gradually varied, unsteady flow (i.e., the Saint Venant flow equations). Considering the different type of flows that have to be represented in the network, we select the Dynamic Wave routing that produces the most theoretically accurate results. The model can not be calibrated because we are representing a non existent sewer network, therefore there is no data again with which to calibrate or validate the model. One option is to compare the peak of the hydrograph with the discharge estimation of the steady state solution of the design (WWF=8968 l/s). From the hydrographs shown in Figure 4.12d it seems that the best representation of the peak flow of the catchment is obtained with the Chicago method, therefore we decided to continue the analysis of the sewer with this method. In this case, the peak of the hydrograph is approximately 400 l/s bigger than the peak estimated in the design. Even though, there are differences, the model appears to represent properly the hydraulic design of the system.

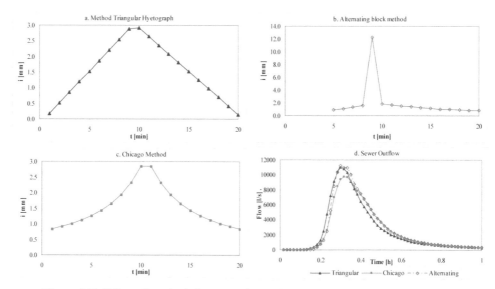

Figure 4.12 Effect of synthetic hyetographs on the sewer hydrograph at the outflow

Modelling water quality in the sewer network

The main aim of the water quality modelling is to simulate the variation in the concentration of pollutants with time at the outlets of the sewer system (i.e. CSO and flow to WwTP). These simulations will help to understand the effect of the design variables on the performance of the system. For the case study, the performance is associated with the water quality requirements of the Lili River. As stated above the main concern of the environmental authority is to maintain the concentration of DO, BOD and TSS above the standards. In addition, models of wastewater treatment plants are COD based and a proper balance of the nutrient processes is required. Considering that BOD can be represented by a fraction of COD and that DO in a sewer can be neglected we select three parameters to be modelled: TSS, COD and nitrogen as Total Kjeldahl Nitrogen (TKN).

SWMM can simulate the generation, inflow and transport of any user-defined pollutants. Pollutant build-up and wash-off from sub-catchment areas are determined by the land uses assigned to these areas. In the model for Sector 1A we used two land use types: residential sub-urban and lawns and parks. Input loadings of pollutants to the drainage system can also originate from external time series inflows as well as from dry weather inflows. For the DWF we include the average values for the year 2006 measured at Cañaveralejo WwTP: 196 mg/l, 482mg/l and 60 mg/l for TSS, COD and TKN respectively.

The build-up (accumulation) of pollutants in the catchment is a function of the number of preceding dry weather days. SWMM has three alternative functions to model the accumulation: power, exponential or saturation. Pollutant build-up that accumulates within a land use category is normalized by either a mass per unit of sub-catchment area or per unit of curb length. From these alternatives, we selected the exponential function (Eq. 4.1) normalized by the area of the sub-catchment. The exponential function was used to estimate the build-up of dust and dirt accumulated in the catchment area. For each particular pollutant, the build-up (B) was estimated as a fraction of dust and dirt.

$$B = C_1(1 - e^{C_2 t})$$

Eq. 4.1

Where: C_1 = maximum build-up possible (mass per unit of area) and
C_2 = build-up rate constant (1/days).

The wash-off of pollutants from a given landuse category occurs during wet weather periods and can be described by an exponential function, a rating curve or event mean concentrations. From these alternatives, we select the exponential function to estimate the wash-off (W) of each particular pollutant (Eq.4.2).

$$W = C_3 q^{C_4} B$$

Eq. 4.2

Where: C_3: wash-off coefficient,
C_4: wash-off exponent,
q : runoff rate per unit area (mm/hour), and
B : pollutant build-up in mass units.

The parameters of the functions for build-up and wash-off can be used to calibrate the model. Because there is no data to calibrate it, we select values based on an application of SWMM model for a combined sewer system in Spain (Temprano, *et al.* 2005). Parameters were slightly modified according to the combination that generates peaks of pollution with the order of magnitude measured as maximums at Cañaveralejo WwTP (see Figure 4.11 above).

Table 4.6 Parameters of the equations of build-up and wash-off

Function	Parameter	Value	Unit
Build-up Exp	Maximum build up C1	12	kg/ha/d
	Build-up rate constant C2	0.3	1/d
Wash-off Exp	Wash-off coefficient C3	2.5	1/mm
	Wash-off exponent C4	1	
Pollutants	Factor for TSS	1	g/g
	Fraction of co-pollutant COD/TSS	1.5	g/g
	Fraction of co-pollutant TKN/TSS	0.08	g/g

4.5.2   Wastewater treatment plant model

In this section the implementation of the model for the wastewater treatment plant of Sector 1A -Cali is presented. The objective of the model is to understand the effect of the design variables of the sewer system on the performance of the wastewater treatment and specifically the variation of the effluent quantity and quality. Therefore, there is no need for a detailed model of all processes in the treatment plant. The focus was only on the main process of the liquid phase. The process for the solid phase, like sludge dewatering or sludge decomposition are not included in the model. A time step of 5 minutes was used to be able to describe the variation in the effluent's quantity and quality.

Schematization

The model describes the activated sludge and clarification processes. The activated sludge processes are represented using the ASM1 model implemented in MATLAB by the author. Five combined steered tank reactors (CSTR) in series are included for that purpose. The anoxic zone is represented by two CSTR and the aerated reactor is represented by three CSTR. The secondary settler is modelled using the Tackas model without biological reactions, implemented in MATLAB by the author. The settler is subdivided in 10 layers and the feed point is located in the sixth layer. Two return flows are included: the nitrate internal recycle (Qir) from the 5th to the 1st reactor and the return of activated sludge (Qrs) from the underflow of the secondary settler to the front end of the plant. The excess of activated sludge is wasted continuously from the secondary settler underflow at a rate Qex. Only one lane is modelled. The schematization of the treatment plant model is presented in Figure 4.13.

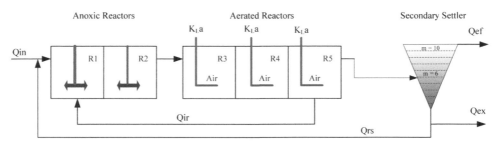

Figure 4.13 Schematization of the Wastewater Treatment Plant Model for Sector 1A

Note: Qin: influent flow, Qir: internal recycle, Qex: sludge wasted, Qrs: return sludge, Qef: effluent flows, K_La: oxygen transfer coefficient. R1 to R5: reactors.

The dimensions of the treatment plant components are presented in Table 4.7. The total volume estimated in the design is equally distributed between the reactors. The total depth of the secondary settler is also equally distributed in the ten layers used in the model to describe the settling process.

Table 4.7 Dimensions of reactors used in the model of the treatment plant of Sector 1A –Cali

Component	Dimension/reactor	Value	Unit
	Volume	1000	m^3
Anoxic reactor R1 and R2	Depth	4	m
	Area	250	m^2
	Volume	2375	m^3
Aerated reactor R3, R4 and R5	Depth	4	m
	Area	594	m^2
	Volume	1728	m^3
	Depth	4	m
Secondary settler	Area	432	m^2
	Number of layers	10	
	Feed Layer	6	

Characterization of flows and wastewater composition
The influent flows and composition for WwTP model are generated with the sewer model. In dry weather the influent flow (Qin) corresponds to the DWF estimated for the design of the sewer. For wet weather the flows depend on the rainfall hyetograph modelled in the sewer. The fractions of return sludge, internal recycle and excess sludge wasted are adjusted in order to achieve a solid retention time (SRT) in the order of 20 days for dry weather conditions. The oxygen transfer coefficients are selected based on having a dissolved oxygen concentration in the last aerated reactor below 1 mg/l. Flow rates and oxygen transfer coefficients used in the model are listed in Table 4.8.

Table 4.8 Flow rate variables and oxygen transfer coefficients

System variables	Value	Unit
Influent flow rate (Qin = DWF)	6480	m³/d
Return sludge fraction (RSf)	0.6	
Internal recycle fraction (IRf)	3.85	
Return sludge flow rate (Qrs)	3888	m³/d
Internal recycle flow rate (Qir)	24948	m³/d
Wastage flow rate (Qex)	150	m³/d
Oxygen transfer coeffcient K_La R3	140	1/ d
Oxygen transfer coeffcient K_La R4	140	1/ d
Oxygen transfer coeffcient K_La R5	25	1/ d
Saturated oxygen concentration	8	mg/l

The ASM1 has 13 components (state variables). Table 4.9 lists the ASM1 state variables, the associated symbols, the values for dry weather in Cali and the units. The values for SS; XB,H; XS; XI; SNH; SI; SND; XND are estimated as a fraction of the average concentration of TSS, COD and TKN used for the design of the treatment plant. The values for SO, XP; and SNO are assumed to be zero and SALK is given a default value of 4 mol/L.

Table 4.9 Influent composition for dry weather flows.

State Variable Description	State Symbol	Value for dry weather	Units
Soluble inert organic matter	SI	33.7	g COD m-3
Readily biodegradable substrate	SS	120.5	g COD m-3
Particulate inert organic matter	XI	96.4	g COD m-3
Slowly biodegradable substrate	XS	231.4	g COD m-3
Active heterotrophic biomass	XB,H	0.05	g COD m-3
Active autotrophic biomass	XB,A	0.05	g COD m-3
Particulate products arising from biomass decay	XP	0	g COD m-3
Oxygen	SO	0	g COD m-3
Nitrate and nitrite nitrogen	SNO	0	g N m-3
NH4 + + NH3 nitrogen	SNH	42.0	g N m-3
Soluble biodegradable organic nitrogen	SND	3.0	g N m-3
Particulate biodegradable organic nitrogen	XND	15.0	g N m-3
Alkalinity	SALK	4.0	mol L-1

Steady state simulation

Because the model is used to test the design of an UWwS, is not possible to calibrate the model. However, for a dynamic use of the model, it is necessary to establish the initial conditions of the variables of the system in each reactor (in total the model has 153 state variables). This ensures a consistent starting point and minimizes the influence of starting conditions on the generated dynamic output. The initial conditions are defined as the state of the variables for steady state conditions. Therefore, the first step in the simulation procedure is to simulate the system under study to steady state using an influent of constant flow and composition. Steady state is defined by simulating 100 days using a constant influent. The dry weather data presented in Table 4.8 and Table 4.9 are used for this purpose. The stoichiometric and kinematic parameters used in model for the activated sludge process and the settling parameters are presented in Table 4.10.

Table 4.10 Model parameters for wastewater treatment plant

Model parameters	Symbol	Value	Unit
Stoichiometric parameters			
Heterotrophic yield	YH	0.67	g cell COD formed (g COD oxidized)-1
Autotrophic yield	YA	0.24	g cell COD formed (g N oxidized)-1
Fraction of biomass yielding particulate products	fP	0.08	dimensionless
Mass N/mass COD in biomass	iXB	0.08	g N (g COD)-1 in biomass
Mass N/mass COD in products from biomass	iXP	0.06	g N (gCOD)-1 in endogenous mass
Kinetic parameters			
Heterotrophic max. specific growth rate	μ_H	4	day-1
Heterotrophic decay rate	b_H	0.3	day-1
Half-saturation coefficient (hsc) for heterotrophs	K_S	10	gCOD/l
Oxygen hsc for heterotrophs	K_{OH}	0.2	gCOD/l
Nitrate hsc for denitrifying heterotrophs	K_{NO}	0.5	g NO3-N m-3
Autotrophic max. specific growth rate	μ_A	0.5	day-1
Autotrophic decay rate	b_A	0.05	day-1
Oxygen hsc for autotrophs	K_{OA}	0.4	g O2 m-3
Ammonia hsc for autotrophs	K_{NH}	1	g NH3-N m-3
Correction factor for anoxic growth of heterotrophs	η_g	0.8	dimensionless
Ammonification rate	k_a	0.05	m3 (g COD day)-1
Max. specific hydrolysis rate	k_h	3	g slowly biodeg. COD (g cell COD day)-1
Hsc for hydrolysis of slowly biodeg. substrate	K_X	0.1	g slowly biodeg. COD (g cell COD)-1
Correction factor for anoxic hydrolysis	η_h	0.8	dimensionless
Multi-layer Double Exponential Settling Velocity			
Maximum settling velocity	Vs	250	m.d-1
Maximum Vesilind settling velocity	Vso	474	m.d-1
Hindered zone settling parameter	rh	0.000576	m3.(g SS)-1=mg/l
Flocculant zone settling parameter	rp	0.00286	m3.(g SS)-1=mg/l
Non-settleable fraction	fns	0.00228	dimensionless

The results of the steady state simulation are presented in Figure 4.14. The flows are presented in Figure 4.14a, the internal recycle and the return sludge increase for approximately 5.5 times the flow that passes through the treatment plant. The wastewater composition in the anoxic and aerated reactors and the effluent are presented in Figure 4.14 b, c and d respectively. Steady conditions are reached approximately after 50 days of simulation. The initial conditions for dynamic simulations are selected as the average value of the variables obtain with the model. A summary of selected state variables for the reactors, the effluent and the underflow of the secondary settler is presented in Table 8.6 and the TSS per each layer of the secondary settler is presented in Table 8.7 (See Appendix 8.2).

Composite variable calculations
The results of the WwTP model at the effluent have to be transformed to state variables that can be used in the model of the River Lili. Therefore, composite state variables are calculated using the following equations (Copp 2002):

$$TSSe = 0.75 \ (XS,e + XBH,e + XBA,e + XP,e + XND,e + XI,e) \qquad \text{Eq. 4.3}$$
$$CODe = SS,e + SI,e + XS,e + XBH,e + XBA,e + XP,e + XI,e \qquad \text{Eq. 4.4}$$
$$BODe = 0.25 \ (SS,e + XS,e + (1 - fp) \ (XBH,e + XBA,e)) \qquad \text{Eq. 4.5}$$
$$TKNe = SNH,e + SND,e + XND,e + iXB \ (XBH,e + XBA,e) + iXP \ (XP,e + XI,e) \quad \text{Eq. 4.6}$$
$$NOe = SNO,e \qquad \text{Eq. 4.7}$$
$$Ntote = TKNe + NOe \qquad \text{Eq. 4.8}$$

Note:
The variables are described in Table 4.9 and Table 4.10. The addition of "e" at the end of the ASM1 notation for the state variables indicates effluent composition.

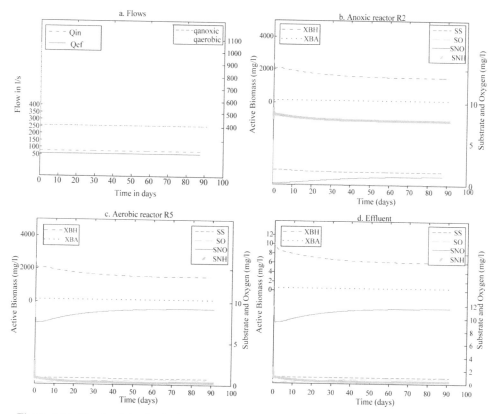

Figure 4.14 Steady state simulations for the wastewater treatment plant of Sector 1A – Cali.

4.5.3   River Lili model

In this section we introduce the model of the Lili River. The main objective of the model is to assess the pollution impacts in the river caused by discharges from the UWwS of Sector 1A - Cali. These imply that the focus of the model is on processes related with the transport and transformation of suspended solids, organic matter, nutrients and their influence on the concentration of oxygen in the river. Even though, one of the problems identified with the receiving system is the lack of capacity of the river to transport peak flows, we assume that the conveyance of the river will be increased as proposed by the consultancy company in the feasibility study (EMCALI and Hidro-Occidente SA 2006). Therefore, the analysis is centred on the water quality components in the river reach of interest. Thus the modelling software selected, namely the Water Quality Analysis Simulation Program (EPA WASP 7.3) has sufficient complexity to simulate the transport and the transformation of the water quality components of interest (Ambrose, *et al.* 1993). A time step of 5 minutes was used to be able to describe the variation of river quantity and quality following the time step used in the sewer and the WwTP models.

Schematization

Lili River is the receiving water for both the effluent of the treatment plant and the CSO. The reach of interest for the case study corresponds to the last 2.95 km of the Lili River, from Calle 48 (S0) to the discharge point in the South Canal (S4), as illustrated in Figure 4.15. The reach length was kept to the minimum to limit the time required to model the system. The state of the river upstream of the starting point of the analysis is included as a boundary condition. The geometry of the river in the segments is based on surveyed cross sections presented in Figure 4.16.

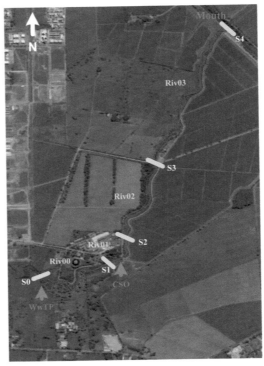

Figure 4.15 Segmentation of Lili River for the model

Information for the schematization and the boundaries of the river is based on the feasibility report for the water supply and sanitation of the expansion zone of Cali (EMCALI and Hidro-Occidente SA 2006) and the analysis of pollutant discharges and water quality objectives carried out by the local environmental authority (DAGMA 2006, DAGMA and UNIVALLE 2009). The characteristics of the segments of the river are presented in Table 4.11.

Table 4.11 Characteristics of the segments of Lili River

Segment Name	Length [m]	Slope [m/m]	Hydraulic Radius[1] [m]	Width[2] [m]	Roughness[3]
Riv00	763	0.0026	0.35	2.95	0.04
Riv01	240	0.0002	0.46	4.50	0.04
Riv02	640	0.0047	0.27	4.86	0.04
Riv03	1300	0.0032	0.63	2.40	0.04

[1,2] Hydraulic radius and width under average flow conditions
[3] Manning coefficients assumed from previous modelling experience of Lili River (EMCALI and Hidro-Occidente SA 2006)

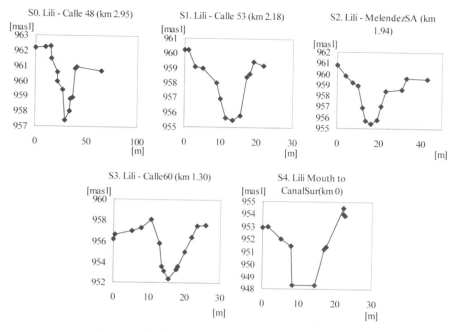

Figure 4.16 Cross sections of Lili River used in the model

Source: (EMCALI and Hidro-Occidente SA 2006)

Modelling flows

WASP is capable of internally calculating flow through a one-dimensional, branching network using the kinematic wave or the fully dynamic formulations (Ambrose and Wool 2009). As for the case study there is no need to analyze backwater effects in the river; the kinematic wave formulation is good enough to simulate the flows in the river reaches of interest. Kinematic flow is controlled by bottom slope and bottom roughness. In WASP a set of user-specified hydraulic discharge coefficients defines the relationship between velocity, depth, and stream flow in surface water segments. This method follows the implementation in QUAL2E (Brown and Barnwell 1987). Discharge coefficients giving depth and velocity from stream flow are based on empirical observations of the stream flow relationship with velocity and depth. The equations relate velocity, channel width, and depth to stream flow through power functions:

$$v = vmult\ Q^{vexp}$$ 　　　　Eq. 4.9
$$R = dmult\ Q^{dxp}$$ 　　　　Eq. 4.10
$$B = bmult\ Q^{bexp}$$ 　　　　Eq. 4.11

Where:　Q: flow [m3/s]
　　　　v: velocity [m/sec]
　　　　R: hydraulic radius, or cross-sectional average depth [m],
　　　　B: top width [m]
　　　　vmult, dmult, and bmult: empirical coefficients, and
　　　　vexp, dxp, and bexp: empirical exponents.

The cross-sectional area, A is the product of top width and average depth, and from continuity, the flow is given by:

$$Q = v*A = v*R*B = (vmult * bmult * dmult)Q^{(vexp + dxp + bexp)}$$ 　　Eq. 4.12

71

Therefore, the following hydraulic relationships hold:

$$vmult * bmult * dmult = 1 \qquad \text{Eq. 4.13}$$
$$vexp + dxp + bexp = 1 \qquad \text{Eq. 4.14}$$

For site-specific river or stream simulations, hydraulic coefficients and exponents must be estimated. In WASP, the Kinematic Wave Flow option requires the specification of hydraulic depth exponent dxp, depth (D) and the width (B) under average flow conditions. The hydraulic geometric equations and Manning's equation are used to calculate the rest of hydraulic coefficients and exponents. For the River Lili, there is no information available of the segments hydraulics to derive the coefficients and exponents. Therefore, we use a one dimensional, fully dynamic model of Lili River implemented in MIKE11 to simulate a range of flows. Based on the results of depths and flows, we fit a power functions for each segment. The resulting functions are shown in what follows:

$$\text{Riv01} \quad D = 0.669*Q^{0.3957} \quad R^2 = 0.9989 \qquad \text{Eq. 4.15}$$
$$\text{Riv02} \quad D = 0.418*Q^{0.4515} \quad R^2 = 0.9985 \qquad \text{Eq. 4.16}$$
$$\text{Riv03} \quad D = 0.745*Q^{0.1774} \quad R^2 = 0.9903 \qquad \text{Eq. 4.17}$$

The initial depth exponent coefficients (dexp) were selected from equations 4.15 to 4.17, while the average depth and width were also estimated from the results of MIKE11. Finally, a consistent set of hydraulic multipliers and exponents were derived from mean flow width, the hydraulic geometry equations, and Manning's equation. The coefficients and exponents used for the River Lili segments are presented in Table 4.12

Table 4.12 Hydraulic coefficients and exponents for segments of Lili River

Segment Name	dexp	vexp	bexp	bmult	vmult	dmult
Riv00	0.40	0.27	0.33	4.05	0.66	0.37
Riv01	0.40	0.26	0.34	6.22	0.26	0.62
Riv02	0.45	0.30	0.25	6.15	0.67	0.24
Riv03	0.18	0.12	0.70	4.69	0.66	0.32

It should be notice that the Lili River model should be calibrated against measured data. Due to the lack of data for calibration, the results of the Lili model using WASP were compared with a fully dynamic model developed in MIKE 11. Despite the fact that the comparison made is only for consistency of model structures, the results are consistent with what is expected and we assume that the un-calibrated model is representative of the main flow conditions in the River Lili. Because of segment velocities, widths, and depths are an integral part of the pollutant transport simulations, in practical applications the river model should be calibrated and validated with measured data.

Modelling Water Quality
WASP is a dynamic compartment-modelling program for aquatic systems, which has the capability to simulate both the water column and the underlying benthos processes. Time-varying processes of advection, dispersion, point and diffuse mass loading and boundary exchange can all be represented in WASP. Water quality processes are simulated in special modules that can be chosen from a library according to the specific problem.

For the Lili River the water quality parameters to be modelled are included in the module of WASP called EUTRO that is specially developed for a conventional analysis of eutrophication processes. Within the eutrophication model there are six levels of complexity that can be chosen. The selection of the level of complexity for the Lili River is based on the processes that need to be described to simulate the variables of interest for the Lili River standards (DO, BOD_5 and TSS). Thus, the main objective of the water quality model is to properly describe the dissolved oxygen balance and suspended solids. The level of complexity selected is 2, which is called the Modified Streeter-Phelps. The modified Streeter-Phelps equations divide the biochemical oxygen demand into carbonaceous (CBOD) and nitrogenous fractions (NBOD), and allow time-variable temperatures to be specified. The processes that may be included are: re-aeration, carbonaceous de-oxygenation, nitrogenous de-oxygenation, settling of organic material and sediment oxygen demand (Wool, *et al.* 2005).

For the model of the Lili River, CBOD is expressed by BOD_5 and NBOD is expressed by TKN. Therefore, the state variables included in the model are: DO, BOD_5, TKN for oxygen balance and we add TSS for the balance of suspended solids. Settling of organic material and sediment demand were not included in the model of the Lili River. However, level of complexity and processes selected, allows a realistic representation of pollution impacts in the River Lili. Higher complexity will increase the demand for data that is not available, and might not increase the information of pollutants of interest for this case study.

The constants and kinetic parameters for the processes included are presented in Table 4.13. For the re-aeration rate, Owens formula is automatically selected for segments with depth less than 0.6 m. For segments deeper than that, O'Connor-Dobbins are selected based on considerations of depth and velocity (Wool, *et al.* 2005).

The decomposition of carbonaceous organic matter (CBOD) is considered as a first order reaction and therefore is a function of the degradation rate (k_1). The range of values recommended for this rate varies between 0.1 and 1.5 d^{-1} as referenced in Jørgensen and Bendoricchio (2001). Roesner, *et al* (1981) indicate that this ratio is spatially variable and may range between 0.1 and 2 d^{-1}. Chapra (1997) report values of k_1 measured in laboratory at 20°C: for untreated water values range from 0.2 to 0.5 d^{-1} while for the effluent of activated sludge treatment values range from 0.05 to 0.1 d^{-1}. From the analysis of removal rates he concluded that degradation is faster for untreated water than for effluents of WwTP, because treatment removes the part of the organic matter that is easily degraded leaving compounds more difficult to degrade. It also found that the degradation rate tends to be higher immediately downstream of the discharge and suggests that this effect is more pronounced when raw sewage is discharged. The rate of nitrification (k_n) is the rate of consumption of the NBOD matter. According to Chapra (1997), values of k_n generally range from 0.1 to 0.5 d^{-1} for deeper waters, and for shallower streams, values greater than 1 d^{-1} are often encountered. The nitrification and BOD decay rates and temperature coefficients are selected based on the author's previous experiences of modelling the Cauca River and its tributaries (Vélez, *et al.* 2006).

Table 4.13 Constants and kinetics for water quality parameters

Constants and kinetics	Value	Unit
Calc Reaeration Option (0=Covar, 1=O'Connor, 2=Owens, 3=Churchill, 4=Tsivoglou)	1	
Elevation above Sea Level used for DO Saturation	1000	msnl
Theta -- Reaeration Teperature Correction	1.03	
BOD Decay Rate Constant @20 °C (k_1)	1.2	d^{-1}
BOD Decay Rate Teperature Correction Coefficient	1.07	
Nitrification Rate Constant @20 °C (k_n)	1	d^{-1}
Nitrification Teperature Coefficient	1.07	

Boundary conditions

In the analysis of pollution impacts, the upstream boundary conditions may have a significant influence. The current water quality conditions of the Lili River are critical; in consequence, the selection of boundary conditions influences possible pollution impacts. In terms of upstream flows, the effect will be associated with the dilution of pollution in the river; in other words, selecting an upstream flow with a high return period (5, 10 or 20 years) will increase the assimilation capacity of the river, which is not the most critical situation. The effect of different flow boundary conditions is shown in Figure 8.3 (Appendix 8.2). Usually, in waste load allocations, low-flow conditions are used in order to analyze the most critical condition for the receiving system (low auto-depuration capacity). Therefore, for the case study we use the 50% permanent flow condition as the average flow of the river. In addition, for the water quality variables, the values are based on the information reported by the local environmental authority (DAGMA and UNIVALLE 2009). Table 4.14 includes the upstream boundary conditions used for the Lili River.

Table 4.14 Upstream boundary conditions for Lili River

Parameter	Symbol	Value	Units
Total kjeldhal nitrogen	TKN	3	mg/l
Biochemical Oxygen Demand	BOD_5	3.7	mg/l
Dissolved Oxygen	DO	5	mg/l
Total Suspended Solids	TSS	10	mg/l
Temperature	Temp	24	oC
River average flow	Q	0.386	m3/s

Figure 4.17 shows an example of the results of the River Lili model, when a load of pollution is discharged from the urban catchment. The variations of the water quality indicators are as expected for the system being modelled.

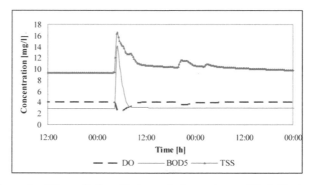

Figure 4.17 Example of the pollution discharge in the water quality indicators of River Lili.

Integrated Model Uncertainty

Each of the models developed here has its own uncertainties. In the case of the models for the sewer and the WwTP, because they cannot be calibrated against measured data, the uncertainties are mainly associated with the model structure and the parameters used. The input data is not an issue since the data used comes from the design of the system. The river model has uncertainties associated with all the main sources: structure of the model, calibration parameters and input data. In addition, the uncertainty may be propagated through the three models. For the case study in hand, the information available is limited to a detailed estimation of the uncertainty. In addition, the focus of this research is not in the uncertainty of integrated models. More on this subject can be found in Freni, *et al.* (2009). However, we recognize that uncertainty may hinder possible benefits for design of UWwS when integrated models are used to analyze alternatives. Therefore, in practical applications the uncertainty and the propagation of the uncertainty should be a matter of detailed analysis. For this hypothetical case study, we assume that the integrated model is representative of the performance of the system.

Once the integrated model is in place, we can use it to assess the performance of the different design alternatives. In what follows, we use the model to evaluate the performance of the pre-designed system and set the bases for the comparison with optimized alternatives developed in subsequent sections.

4.6 Performance of the Pre-Design Urban Wastewater System

To evaluate the performance of the pre-designed UWwS, we use the integrated model of the system. The model of the sewer was initially fed with information for the pipes as calculated using the rational method, and the model of the wastewater treatment was set up with the pre-design data. The main disturbance of the system is a time series of precipitation. The time series includes three days of dry weather, followed by the design rainfall event of 2 year return period and 20 minutes duration and a final three dry days to left the system flush (almost) completely.

The results of the integrated model are presented in Figure 4.18 to Figure 4.20. A selection of model results is presented for the sewer system. Figure 4.18 a) zooms-in on the wet period of the precipitation time series and the hydrograph for sewer pipe # 25 (the last pipe in the network). The peak of the hydrograph (9355 l/s) approximates the peak design flow (8893 l/s) estimated with rational method. The on-line tank implies that the last part of the sewer network will be surcharged during wet weather conditions, modifying the flow in the last pipe. Figure 4.18 b) shows the dynamic of the system in terms of water quality. TSS and COD are used to show the variability of the pollution during the rainfall event. The peak of pollution is generated in the system by the exponential wash off model and can be associated with the flushing of the catchment caused by the first part of the runoff. Figure 4.18 c) shows the hydrograph of the water pumped to the treatment plant during wet weather conditions. This figure corresponds to a set point of the pump equal to 5 times the DWF (that is Qmax= 420 l/s). Figure 4.18 d) shows the hydrograph of the CSO with Qmax= 7668 l/s. The CSO hydrograph is the system response to the combination of two variables: the CSO weir depth, the on-line storage capacity and the set point ratio of the WwTP. For the pre-design scenario the storage volume is 2603 m^3 as a result of using Formula A.

Figure 4.19 shows the response of the wastewater treatment reactors to the variation of the flow and water quality components. Figure 4.19 a) shows flows in the reactors. The inflow to WwTP during the precipitation event is increased 5 times and this may cause a hydraulic disturbance in the system. Figure 4.19a shows how in the model that increment in the inflow is propagated to the anoxic and aerated reactors in the treatment plant. The hydraulic disturbance is exaggerated by the return flows. This is because the internal recycle (Qir) and the return sludge (Qrs) are adjusted according to the influent flow. This two return flows increase 5 times the flow that pass through the reactors.

Figure 4.19 b and c) show the water quality components in the anoxic and aerobic reactors respectively. This figures shows how the wastewater composition is affected by the fluctuations in the influent associated with the precipitation event. The disturbance in the reactors can be seen as sudden increments on some of the parameters like the concentrations of organic matter (SS) and nutrients (SNO) in the Anoxic reactor and a reduction on (SNH) probably caused by dilution of the concentration of Ammonia in the influent (Figure 4.19b). The disturbance is propagated to the aerated reactor and the effluent. In the effluent the nutrient concentrations show peaks that can be critical for the receiving system (Figure 4.19 d). Notice also that the 20 rainfall event generate a disturbance in the system such that require 24 hours to recover to the dry weather operational conditions. The possibility to describe those peaks in the effluent gives importance to the coupling of the sewer model with the WwTP model.

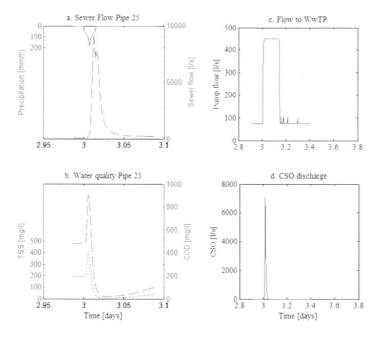

Figure 4.18 Integrated model results: a, b) Sewer pipe 25, c) Outflow to WwTP and d) CSO.

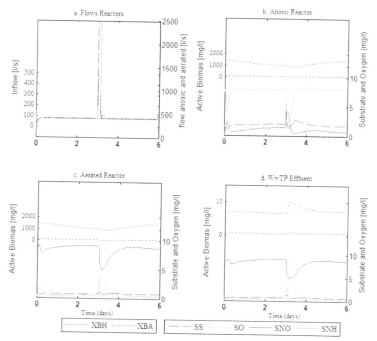

Figure 4.19 Integrated model results: a. WwTP flows b. and c. Reactors and d. Effluent

The output of the CSO from the sewer and the effluent from the treatment plant are the disturbance inputs to the river model. Results for the main water quality components in the last segment of the river are presented in Figure 4.20. From Figure 4.20 a) it is possible to observe the peak of the TSS as a consequence of the discharge of suspended solids from the sewer catchment. During dry weather there is no effect on the TSS due to discharges from the treatment plant such that the concentration in the river staying below the standard set by the environmental authority (TSS <10 mg/l). However, during the rainfall event, the peak of TSS reaches a concentration of 40.8 mg/l for about half an hour. Similar responses are followed by COD and TKN with peaks of 42.8 mg/l and 5.2 mg/l respectively (Figure 4.20 b and c). The oxygen dissolved is perhaps the most critical component because of its acute effect on the river ecosystem (Figure 4.20 d). The minimum concentration of DO during the rainfall event is 1.8 mg/l and the concentration is below the standard 4 mg/l for about 2.3 hours.

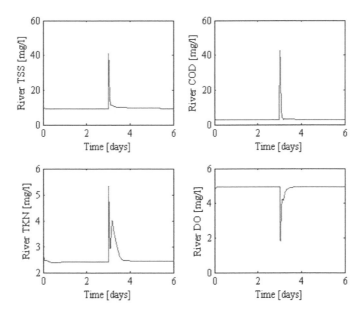

Figure 4.20 Integrated model results: River water quality at the last segment (Riv03).

4.6.1 Definition of the performance indicators

Performance indicators are defined based on the objectives of the case study. Therefore, each objective has an indicator: pollution impacts on the river, flooding in the sewer catchment and the costs of the pipe network and the storage.

Cost of the sewer network and storage
Even though cost is the main driving force for optimizing a sewer network, according to Guo, *et al.* (2008) cost functions are often simplified, generalized or unrealistic, generating solutions that may not be optimum in reality. To increase the reliability of the optimization a cost function was deduced for the specific case study. The main components of the cost are: supply, transport and installation of the pipes, trench excavation and back filling, manholes and storage cost. Cost depends highly on the material used and the local unitary costs. For the pipes we use concrete pipes with diameters that range from 0.15m to 0.53 m and reinforced concrete pipes in the range 0.61 m to 4.05 m. The catalogue is included in Table 8.8 (Appendix 8.2).

The cost per meter of pipe installed is based on the unitary cost used by the utility company for tenders (EMCALI 2004). The prices were updated to 2010 using an inflation of 24.7%, which corresponds to the sum of the annual inflation rates of the Colombian economy from 2004 to 2010, as reported by the Departamento Administrativo Nacional de Estadisticas (DANE). The cost function of pipe installation per meter is shown in (Figure 4.21 a). The cost of excavation and backfill depends also on the depth that the pipes are laid; therefore three different cost functions were created as shown in Figure 4.21 c. Similarly, the cost function of manholes is dependent of the depth of the manhole as shown in Figure 4.21 b. The cost of the storage is based on cost of tenders for the city of Cali and the cost function of storage given by U.S. EPA for storm water control (Heaney, *et al.* 2002). The storage cost as a function of the volume is shown in Figure 4.21 d.

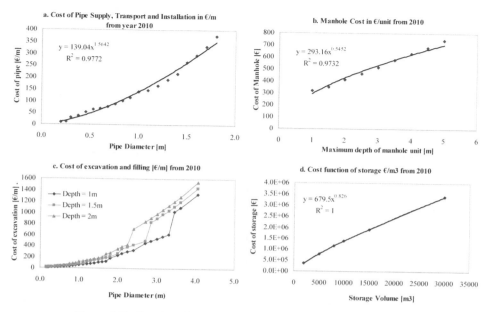

Figure 4.21 Construction cost of sewer network and storage for Cali

Flood and pollution indicators

The model of the sewer has the capability to generate the volume of water that exits any of the manholes or storages of the system and to compute the volume of flooding. As an indicator for the flooding, we use the sum of the volumes of water that exit the sewer network via manholes and storage nodes. This indicator has been frequently used in the optimization of sewer design (Guo, *et al.* 2007). The pollution indicator selected is the minimum DO and the maximum TSS estimated by the model for the river in the last segment. The results of the indicators for the pre-designed sewer network are presented in Table 4.15. The last four indicators are used as a base scenario for comparison with alternative designs derived using MoDeCo approach. In general the pre-designed system performance is very good in terms of flooding control but very poor in managing the pollution impacts in the River Lili.

Table 4.15 Performance indicators for the pre-designed sewer network of Sector 1A.

Indicator	Value	Unit
Pipe cost	861413	€
Excavation and backfillingl cost	718500	€
Manholes cost	11453	€
Storage cost	450093	€
Total sewer network and storage	**2032257**	€
Total flood volume	**0**	m^3
Minimum Dissolved Oxygen	**1.80**	mg/l
Maximum Total suspended solids	**40.8**	mg/l

4.7 Optimum Design of the Sewer Network

4.7.1 Definition of the system to be optimized

Here the approach makes use of the multi-objective optimization technique to optimize the pre-designed sewer network. The integrated model is used to assess alternative designs and to estimate the performance indicators. Following the objectives of the case study, the boundaries for the optimization of the design are closed around the sewer network and include: the pipes, the on-line storage volume and the setting of the pumps to the WwTP. Although, the focus is on the design of the sewer network, a wider view is still preserved by including the interactions of the sewer with the WwTP and the River Lili.

4.7.2 Definition of the inputs and outputs of the system

There are two types of inputs: those that we can manipulate in the system (the degrees of freedom) and the ones that we cannot (the disturbances). The degrees of freedom of the sewer network are: the diameters of pipes, slopes (s), storage volume (Sv) and the set point of the pumping station for wet weather condition, which is considered as the ratio of Qp/DWF (R_{Qp}). The disturbance of the system is the time series of precipitation. The main outputs of interest are the state variables of the system that allow us to estimate the performance indicators: flood volume in the sewer network, concentration of water quality components in the last segment of the river modelled (DO and TSS) and the total cost of the sewer network (Cost). Additional outputs of the models can be used to characterize the performance of WwTP, but these are left for the post-processing of the optimum alternatives.

4.7.3 Definition of the objective function and constraints

The optimization of the sewer network can be posed as a multi-objective optimization in which the aim is to find the combination of pipe diameters, storage volume and pumping flow that minimize the flooding, the pollution impacts and the cost of the system. Mathematically the problem can be stated as follows:

$$MinF(x) = \{f_{TFlood}(x), f_{Tpollution}(x), f_{TCost}(x)\}$$

$$Where \; x = [\phi_1 ... \phi_{np}, Sv, R_{Qp}]$$

and x is boundary constrained

$$LB \leq x \leq UB \qquad\qquad\qquad\qquad \text{Eq. 4.18}$$

and each pipe diameter may have inequality constrains

$$\phi_{i+1} \geq \phi_i \; for \; i = [1...np]$$

and ϕ_{i+1} being inmediate downstream pipe of ϕ_1

The objective functions are:

$$TFlood = \sum_{i=1}^{n:nodes} Vol_{Flood\,i} \qquad\qquad \text{Eq. 4.19}$$

$$Tpollution = \begin{cases} TSS \max \\ DOsag = DOsat - DO \min \end{cases} \qquad\qquad \text{Eq. 4.20}$$

$$TCost = \sum_{i=1}^{Pipes} Cu_i L_i + \sum_{j=1}^{Assets} Cu_j A_j$$

Eq. 4.21

The objective functions are based on the performance indicators of the system.
- The flood function corresponds to the sum of volume of water that exits any of the manholes in the system.
- Pollution impacts were measured in comparison with the minimum DO in the river. Trials were carried out using composite indicators that in addition to the minimum DO include the duration (DU) that a standard value is exceeded. The composite pollution function DO-DU proposed by Schütze, et al (2002) was implemented. However, the values obtained with this function were all very close to each other and the optimization algorithm tends to lose the driving force of the pollution impacts. Alternatively, the minimization of the DO sag was implemented using DO saturation (DOsat=8 mg/l) and the minimum DO estimated by the model.
- The cost function corresponding to the sum of the cost of the sewer network and the cost of the storage estimated using the functions derived for the specific case and presented in Figure 4.21.

The decision variables X were defined as follows:
- Pipe sizes were selected from 11 possible options according to discrete sizes from the catalogue (Table 8.8). The upper and lower boundaries of the 11 diameters were defined using as mean size the estimated pipe diameter in the pre-designed network.
- The roughness coefficient was fixed assuming that only one type of pipe was used (concrete).
- The slope (s) was adjusted by an auxiliary algorithm according to the pipe sizes of the network. Thus, a minimum slope per pipe diameter was pre-defined such that the slope guarantees a minimum velocity for self-cleaning and a minimum cover depth as defined by the user (here 1m).
- The storage capacity was included as a decision variable (volume). The lower and upper boundaries of t the volume variable were selected based on the preliminary design of the system.
- The settings for control/operation of the ancillary infrastructure

Even though, the simplifications of the case study, the search space of solutions is huge. Assuming that each pipe has 10 possible diameters, and the storage volume and the pumping ratio are discrete variables with 10 possible steps; the number of possible solutions to the problem is 10^{27}.

4.7.4 Response surfaces for the design of sewer network for Sector 1A

The sewer network design can be seen as a complex multi-objective optimization process. Some of the variables are discrete such as the pipe diameters because they depend on commercial availability of diameters; in contrast the storage volume and the pumping capacity are continuous variables. Due to the great number of variables (27), it is computationally very expensive to make an exhaustive search of all the possible combinations. However, it is possible to assume that all variables are discrete in order to map the response surfaces of the objectives. To do the analysis, a time series of precipitation with the design rainfall event was used. The search space of the pipes was simplified to 11

81

scenarios, varying all the pipe diameters in the network by one size above or below the pre-designed size. The search space for the storage volume (Sv) and the pumping ratio (R_{Qp}) was simplified to only 5 possible steps. Thus, following an exhaustive search, the possible combination of set points is 11*5*5 which means that the model of the system should run 255 times. Even with the coarse grid of pipes, volumes and pipes, the mapping of the response surfaces of 27 variables with three objective functions is very complicated. Therefore, we group the different pipe networks by estimating the total volume of the network. Thus, from 27 variables, we reduce to three variables that can be visualized 2 by 2.

Figure 4.22 shows a sample of the response surfaces generated by the integrated model of the system. The first group of figures a1) to a4) represent the sensitivity of the objective functions to the pipe networks and storage volume. Figure 4.22 a) and b) shows the effect on the objectives of the pollution measured as the minimum DO and maximum TSS respectively. The increment in the size of the pipes seems to have a low impact on pollution control. In contrast, the increment in the storage volume tends to have more impact on the minimum DO and TSS; however, it requires large volumes (Sv>10000 m^3). The other two objectives have a more clear response with respect to the pipe diameters: as pipes are increased the flood is reduced and the cost is increased (Figure 4.22 d and b). It is also clear that there is a minimum pipe network at which the flood is reduced to zero and there is no need to continue increasing pipe size. The storage, on the other hand, plays no role in reducing the flooding volume; but, this is expected because is located at the end of the sewer network, and therefore has no major influence upstream.

The right hand side of Figure 4.22 (b1 to b4) shows the response of the objectives for the pipe network and the pumping flow ratio (R_{Qp}) with a fixed storage volume. Figure 4.22 b1) shows the response surface for minimum DO. The surface shows that increasing the pumping ratio (R_{Qp}) has a positive impact on the dissolved oxygen. This is because more wastewater gets treated, thus reducing the pollution impacts on the river. Notice that the largest DO is obtained with the sewer network witch has the lowest volume (smaller pipes). The explanation for this is that this configuration of network has the biggest flood volume, and the model assumes that the water is ponded and, re-enters the system when the pipe capacity permits it. Therefore, even though, it is a good solution for DO, is very bad for flooding. These contradictory objectives support the methodology proposed and help solve the problem as one multi-objectives optimization. In terms of flooding and cost, the set point of the pumping station has no influence on the response surface, because the pump is at the outlet point to the treatment plant and the pumping cost is not included (Figure 4.22 b3 and b4). In conclusion, the response surfaces show the effects of the selected variables on the objectives and support the design of the sewer network as a multi-variable and multi-objective optimization problem. Thus, the next step is to search for the optimum combination of pipe networks, storage volumes and pumping flows with the support of optimization algorithms.

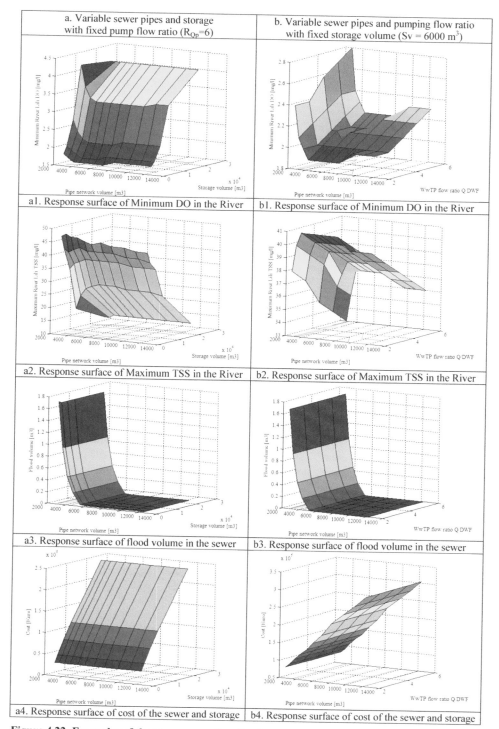

Figure 4.22 Examples of the response surface for the sewer network design of Sector 1A – Cali.

4.8   Setting up the optimization algorithm

The optimization algorithm, as with any model, must be set up. The user of the algorithm must decide which functions such as selection, mutation and crossover are to be included and he must specify the parameters that control the MOEA's search: population size, number of generations, probability of mating, probability of mutation and stopping criteria. According to Nicklow, *et al.* (2010) the run duration should grow proportionally to the number of decision variables being searched; and the average computational time for design evaluations should guide decisions on the population size and run duration. Some effort in the research has been dedicated to define the parameters of the MOEA (Reed, *et al.* 2000); however, recommendations for setting the parameters of the evolutionary algorithm arose from a considerable body of prior analysis on the performance as a function of these parameters. For this case study, the settings are the result of the trial and error process with an analysis of the sensitivity of the parameters for the design of the sewer network. For instance the crossover fraction (probability=0.75) was selected after the sensitivity analysis of the variable for the performance of the optimization as presented in Figure 8.4 (Appendix 8). Different settings were tested and the final results were obtained with the settings presented in each particular optimization exercise.

In particular, Kollat and Reed (2007) have recently shown that these algorithms potentially have a quadratic computational complexity when solving water resources applications. A quadratic complexity implies that a twofold increase in the number of decision variables will yield an eightfold increase in the number of function evaluations required to solve for an application.

4.8.1   Modification of the operators of NSGAII

The standard NSGAII from MatLab is a real number coded algorithm. The optimization of the sewer network for Sector 1A has both discrete and continuous variables: the pipe diameters are indexed as discrete integer values and the storage volume and the pump ratio are continuo variables represented by real numbers. It was necessary to adapt the NSGAII code to generate integer pipe diameters and continuous real values for the other two variables.

In addition, the operators of NSGAII should generate solutions that comply with the inequality constraints on the sewer pipes as described in equation 4.18. There are a number of approaches for constraint handling which include: penalty functions, repair or local search operators, modified mating/mutation functions that preserve constraints, and multi-objective formulations where constraints are reformulated as objectives (Back, *et al.* 2000). For this case study we use both penalty functions and modified creation, crossover and mutation functions.

The optimization algorithm NSGA II uses a random population initialization scheme. That is, all solutions in the initial population are uniformly selected within the boundaries of the variable space (Coello Coello, *et al.* 2002). The crossover and mutation operators do not consider the inequality constraints; therefore, penalty functions had to be included to force the elimination of solutions that do not fulfil the pipe diameter constraint. The first trial using penalty functions shows that many of the solutions were not feasible because they do not comply with the inequality constraints. Figure 4.23 shows the number of feasible solutions in each generation of an optimization exercise using the default operators of NSGAII as coded in MatLab. In the initial population, out of 100 vectors only 10 were feasible solutions. With the evolution of the populations, after 50 generations less than half of the individuals were feasible.

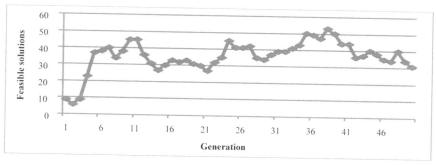

Figure 4.23: Feasible solutions in the evolution of 100 individuals using the default NSGAII functions in MatLab.

The main problem with penalty functions is the definition of good penalty factors that can guide the search towards the feasible region (Coello Coello, *et al.* 2002). Event more, the factors should promote competition between infeasible solutions and should not impact the scaling of the decision space (Nicklow, *et al.* 2010). Figure 4.23 shows that one single factor per objective for all infeasible solutions do not properly handle the search. One alternative is to scale the penalty factor depending on the degree of violation of the constraints. This was not tested during this research.

In order to increase the possibility to have feasible alternatives, the creation, the crossover and the mutation functions we modified. For the creation function, instead of selecting all variables at one time, the elements of the vector that correspond to the diameters were selected one by one according to the location of each pipe and the constraints. The main difference with the default function is that the boundary constraint of the pipes is dynamically updated as the diameter of the pipe upstream is generated. The pseudo-code is given by:

$$if\ \Phi_i\ is\ head\ of\ branch$$
$$\Phi_i = random\ [LB,\ UB]$$
$$else$$
$$\Phi_i = random\ [max(LB,\ \Phi_{i\text{-}1}),\ UB];\ where\ i = 1\ to\ np$$

The crossover function used from MatLab is named "crossover-scattered", also called uniform or random crossover. This function uses two selected vectors (parents) from the population for mating. The new vector (kid) is formed by selecting each variable (gene) from one of the parents. Each gene has an equal chance of coming from either parent. Once the new vector is formed the boundary constraints is check. Two modifications were done to the crossover-scattered function: 1) a check was made of inequality constraints to define the feasibility of the new vector (kid), and 2) an option to do a crossover that instead of doing it per variable (pipe), is done per group of variables (branches) as defined by the user. In the case study, for instance, six branches were defined as shown in Figure 4.24. In addition, a repair function for infeasible branches was included to make the whole vector feasible and increase the chance having a sewer network that complies with the constraints.

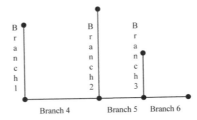

Figure 4.24 Selected branches for modified crossover.

The mutation function available in MatLab uses a single solution to be perturbed in order to generate the new vector (mutant). The function creates the mutated vector using adaptive mutation. By default the new vector satisfies boundary constraints. The function was modified by adding a loop in which the mutation is checked until it also satisfies the inequality constrains of the pipe diameters.

4.8.2 Initial population

One of the known settings that speeds-up the optimization process is the use of known sub-optimal solutions in the initial population. The first known sub-optimal solution corresponds to the pre-design of the system. Therefore in all the optimization exercises the pre-design solution vector was always included as part of the initial population. The use of final populations within an optimization exercise as the initial population for a subsequent optimization was also helpful to find a near optimal set of Pareto solutions.

4.9   Results of the Optimization of the Sewer Network Design

4.9.1   Optimization of pipe diameters using one precipitation event

Characteristics of the optimization
In order to start building the complexity of the model, we decided to optimize only pipe diameters using two objective functions from Equation 4.18: flood volume and cost. The main disturbance of the system was a design rainfall event for a return period of 2 years and duration of 20 minutes. The antecedent conditions were 2 days of dry weather conditions (Figure 4.25). As was observed in the response surface analysis the storage volume and pumping capacity has limited impact on the flood volume; therefore, for this first analysis these variables were fixed using the values of the pre-design.

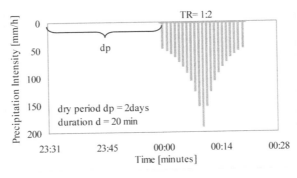

Figure 4.25 Precipitation time series with one rainfall event of 2 years return period

Pareto solutions for the sewer network design

The results of the optimization process are shown in the form of Pareto optimal solutions in Figure 4.26 a). As was expected, the solutions show a trend of increasing flood volume as the costs of the solution found for the algorithm reduces. Figure 4.26 b) shows only the cost of the pipes, by removing the fixed cost of the storage volume. The isolated square point in the figures represents the objective function value for the pre-design system. In general, the results of the optimization show that the algorithm can converge to less expensive solutions with different degrees of flooding. Table 4.16 presents a comparison of different indicators for four solutions selected from both extremes of the Pareto (i.e. low and high flood volume).

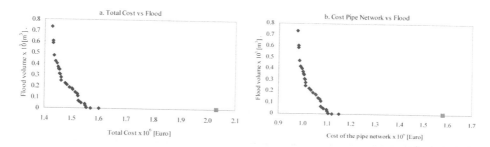

Figure 4.26 Pareto optimal solutions for sewer network of Sector 1A.

Table 4.16 Comparison of optimal Pareto solutions for optimization of pipes

Design Alternative	Pipe Volume m³	Storage m³	Pump ratio Qp/DWF	Flood x10³ m³	DO min mg/l	Pipe cost x10⁶ €	Cost comparison
Pre-design	5977	2603	6.0	0	1.9	1.58	-
1	3399	2603	6.0	0.73	1.8	0.98	-38%
2	3820	2603	6.0	0.06	1.8	1.08	-32%
3	4053	2603	6.0	0.00	1.8	1.12	-29%
4	4207	2603	6.0	0.00	1.8	1.15	-27%

The base scenario for comparison is the pre-designed system. To characterize the alternative designs we use the total volume of the pipes in the sewer network. Alternative 1 has 43% less volume than the base scenario and implies a 38% reduction in the cost of the pipes. However, this alternative may not be acceptable because of the high flood volume generated (730 m³). Alternatives 2 to 4 have a reduction in the pipe diameters that do not cause any flooding in the system for the precipitation event of 2 years TR. In general, we could say that the optimization algorithm may find solutions that are approximately 30% less costly than the base scenario and at the same time maintain the performance of the system in terms of flood and pollution impact at the same level.

System designed with the rational method are normally over-dimensioned (Butler and Davies 2000); thus giving solutions that have extra capacity to deal with more critical rainfall events (TR=20 y) than those they are designed for. Table 4.17 compares the alternative designs and the flood volume for two precipitation events with return periods: TR = 1:2 and 1:20 years. When the solutions found with the optimization process are tested with a more critical precipitation event (i.e. precipitation with TR=20 years), they show as a weakness an increased risk of flooding as a consequence of having smaller pipe diameters. From these results, we may consider including a more extreme rainfall event (i.e. with TR=1:20 years) in the optimization process, if the interest is to maintain the same level of protection from flooding as with the pre-designed system.

Table 4.17 Assessment of alternative designs for a precipitation event of 20 years return period

Design Alternative	Pipe Volume m^3	Storage m^3	Flood TR=2y x10^3 m^3	Flood TR=20y x10^3 m^3
Pre-design	5977	2603	0	0
1	3399	2603	0.73	2.50
2	3820	2603	0.06	1.76
3	4053	2603	0.00	1.42
4	4207	2603	0.00	0.91

In addition, when considering water quality objectives, events that are critical for flooding may not be the most critical events to evaluate pollution impact. On the contrary, more frequent and small volume precipitation events (called micro-storms) are responsible for most of the annual urban runoff discharges. Pitt and Voorhees (2000) classify these rains as follow:

- Events with depths less than 12 mm are associated with low pollutant discharges. However, they may cause water quality violations related with bacteria standards. In most areas, runoff from these rain storms should be totally captured and either re-used or infiltrated in upland areas of the catchment.
- Rains between 12 and 38 mm are responsible for about 75% of the nonpoint source pollutant discharges. These events are key rains in terms of addressing pollutant discharges. Therefore, runoff from these events should be treated, to prevent pollutant discharges from entering the receiving waters.
- Rains greater than 38 mm are associated with the drainage design and are only responsible for relatively small portions of the annual pollutant discharges.

Figure 4.27, shows the results of the pre-designed network when two events of TR=2:1 years (26 mm) and TR=1:20 years (37 mm) are used as disturbance. Even though the concentrations of pollutant generated by the precipitation events are similar, the peak of pollution load is larger for the most frequent event (TR=2:1 year). The extra rainfall in the 20 year RT event may dilute the pollutants and reduce the impact in the receiving system. From these results, we may consider including a more frequent rainfall event (i.e. TR=2:1 year) in the optimization process, if the interest of the optimization is to minimize pollution impacts on the receiving system (Vélez, *et al.* 2009).

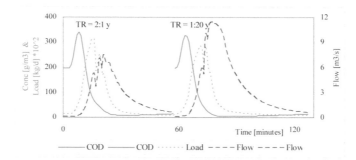

Figure 4.27 Pollutograph and hydrograph of pipe 25 form after two precipitation events.

4.9.2 Optimization of pipe diameters using three precipitation events

Characteristics of the optimization

The main disturbance of the system was set as a time series of precipitation with three rainfall events of TR = 2:1, 1:2 and 1:20 years. The rainfall events have a duration of 20 minutes and the antecedent conditions are set to 2 days of dry weather conditions (Figure 4.28). The optimization variables and objectives are as in the previous optimization. Therefore there are 25 pipe diameters and two objectives: flood volume and cost.

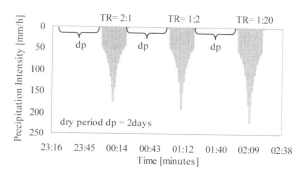

Figure 4.28 Precipitation time series with three rainfall events

For this experiment four different optimization exercises are presented. The variations in the settings of the optimization algorithm for each run are presented in Table 4.18. The objective of these variations was to try to approximate the optimal set of Pareto solutions.

Table 4.18 Settings of the optimization algorithm for Run 1 to Run 4

Optimization	NSGA II Functions	Population	Generations	Stoping criteria	Runing time (hr)
Run 1	Default with penalty function	80	50	Generations exceeded	38.3
Run 2	Default with penalty function	200	127	Average change in the spread of	2.4
Run 3	Default with penalty function	200	102	Pareto solutions $< 10^{-6}$	4.4
Run 4	Modified functions	200	229	Generations exceeded	6.7

Pareto solutions for the sewer network design using three rainfall events

The results of the optimization Run 1 are shown in the form of Pareto optimal solutions in Figure 4.29a). The cost of the pipes versus the flooding volume is shown in Figure 4.29 b). The square point in the figures represents the objective function value for the pre-designed system. As was expected, solutions that have low flood volume are closer to the pre-designed alternative. In other words, the optimization algorithm found solutions that are optimal for the more critical precipitation event.

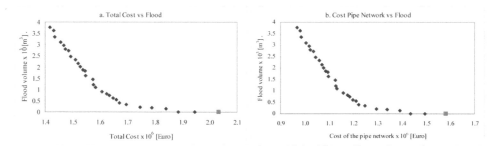

Figure 4.29 Pareto optimal solutions for sewer network of Sector 1A. Optimization run 1.

Table 4.19 presents a comparison of indicators for four solutions selected from both extremes of the Pareto (i.e. low and high flood volume). The alternative 7 found by the optimization algorithm seems to be the best alternative, because, it maintains the same level of protection against flooding for a 20 year TR event as the pre-designed solution, and is 9% less costly. The cost reduction found is consistent with results of the optimization of sewer design found in the literature. Normally the reported values of cost reduction are between 5 and 30% (Guo, et al. 2008). The optimized solution found with the algorithm is similar to the pre-designed solution using the rational method; and the cost savings appear very modest to justify an optimization.

Table 4.19 Comparison of optimal Pareto solutions for sewer pipes of Sector 1A

Design Alternative	Pipe Volume m³	Storage m³	Pump ratio Qp/DWF	Flood x10³ m³	DO min mg/l	Pipe cost x10⁶ €	Cost comparison
Pre-design	5977	2603	6.0	0	1.9	1.58	-
5	3362	2603	6.0	3.76	1.8	0.97	-39%
6	5214	2603	6.0	0.13	1.8	1.39	-12%
7	5421	2603	6.0	0.01	1.9	1.44	-9%
8	5752	2603	6.0	0.00	1.9	1.50	-5%

In an attempt to find a better Pareto set of solutions three more runs of the optimization were carried out with different setting. One of the causes of the low cost saving could be that the Pareto was not close to the optimal set of solutions because of the limited number of function evaluations. In Run 1, the optimization process evaluated 4000 alternatives. The limiting factor was the computing time required. For 4000 evaluations, the computing time in a desktop computer with a CPU capacity of a 3.52 GHz was approximately 40 hours.

In order to reduce the computational time, modifications to the set up of the optimization were carried out: a) the integrated model was decoupled, so that only the sewer network was running, 2) the precipitation time series was simplified to 1 dry period of 6 hours, one precipitation event of 20 minutes with TR=1:20 years and 6 hours after the event of dry weather period. With the simplified model it was possible to target more function evaluations. Experiments with a 29 pipe sewer network reported by Guo, et al. (2007) show that to find solutions near the optimum Pareto set he required 80000 evaluations with NSGAII, and 30000 using an hybrid algorithm named CA-GASiNO.

Thus, for Run 2, the optimization algorithm was set up with a population of 200 vectors, and as stopping criteria the average change in the Pareto Spread with a tolerated limit of $T < 10^{-6}$.

Run 3 used the final Pareto set of Run 2 as its initial population; therefore the number of evaluations is considered as the sum of the two runs. Run 4 targeted changes in the operators of the NSGAII as explained in the section 4.8.1. The final Pareto set of the three runs is presented in Figure 4.30. In addition, the figure includes the Pareto from the Run 1 and the value of the flooding volume function obtained with the pre-designed solution. The results show that in the upper part of the Pareto Front there was a significant improvement when comparing Run 2 with Run 1. However, the best Pareto set is found after 45800 evaluations in Run 3. The Run 4 was set up to compare with the default operators with the modified operator.

Two indicators were used to compare the Pareto sets: the average crowding distance and the Pareto Spread. Table 4.20 shows the indicators for Run 2, 3 and 4. The indicators show that the modified operators approximate the Pareto in a way that is very similar to what was found by the default operators. The main difference is in the distribution of the solution in the Pareto that seems to be better for the Run 4. However, no clear conclusion can be drawn as to which one is better from the exercise. More research is needed to find the possible benefits of modifying the operational functions of NSGAII to handle the inequality constraints.

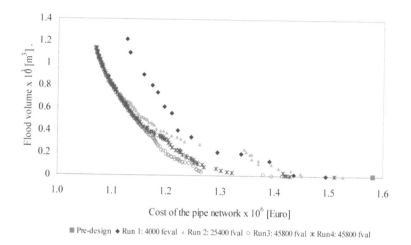

Figure 4.30 Set of Pareto Solutions for four different settings of the optimization algorithm

Table 4.20 Indicators of the Pareto set of solutions.

Optimization	Average Crowding Distance	Average Pareto Spread
Run 2	0.011	0.556
Run 3	0.009	0.516
Run 4	0.008	0.449

From the practical point of view, the area of major interest in the Pareto is the lower part where the flooding function is equal or near to zero. However, it seems that this area of the search space is explored less by the algorithm. Perhaps, the optimization process will benefit from the use of hybrid approaches in which a local search algorithm is coupled with NSGAII. However, the level of detail achieved with the optimization is considered enough for the purpose of the case study. A set of solutions from the Pareto set is post processed and presented in Table 4.21.

Table 4.21 Optimal sewer network of Sector 1A for a flood protection of 1:20 years.

Design Alternative	Pipe Volume m³	Flood x10³ m³	Pipe cost x10⁶ €	Cost comparison	
Pre-design	5977	0	1.58	-	
9	5255	0.00	1.43	-10%	Run 4
10	5219	0.01	1.42	-10%	Run 4
11	5166	0.00	1.41	-11%	Run 3
12	5089	0.00	1.40	-12%	Run 3
13	4894	0.03	1.32	-16%	Run 4
14	4585	0.03	1.27	-20%	Run 3

Six of the sewer network solutions that produce near zero flooding volumes found in the Pareto set from Run 3 and Run 4 are compared with the pre-designed scenario. The smaller pipe volumes correspond to the smaller network formed with the smaller diameters. In general, the cost reduction range between 10 and 20%, which is what, was expected. Alternative 14 from Run 3 seems to be the best solution found by the optimization algorithm. A comparison of the pipe sizes shows that increasing the sizes of a few of the pipes the flooding volume can be reduced to zero, but the cost saving is reduced from 20% to 15%. The pipe diameters of the pre-designed sewer network are compared with the best solution so far in Figure 4.31. The main savings of the cost are achieved by the reduction of the last 5 pipe diameters that are the biggest in the network. That brings to the forefront methodological reasoning: in an optimization exercise of a sewer network, one should include only the main sewers (exclude branches) as the driving variables.

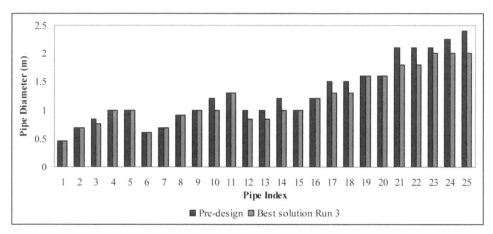

Figure 4.31 Pipe diameters of the sewer network pre-designed and alternative solution 14

Even though, the optimized solution found with the MoDeCo approach is similar to the pre-designed solution using the rational method, and therefore cost savings are the order of 15%, the results of this exercise are valuable. First, the method proves to be capable of find an optimum solutions for a 25 pipe network. Second, the solutions are optimum for the precipitation event specified. Thus, when the decision maker has decided the level of protection required for flooding, the method finds a solution for that specific return period, avoiding the over dimensioned estimations obtained with traditional methods. And third, as the method gives a Pareto set of solutions, the decision maker has more design alternatives from which an informed solution can be made.

As can be observed in Table 4.19, the pollution impact in the river measured as the minimum DO is not improved in any of the alternatives found. Up to now we have designed the sewer network by only considering flooding objectives. In the next section the water quality objectives are included in the optimization process.

4.9.3 Optimization of storage volume and pumping capacity

Characteristics of the optimization
The main disturbance of the system was set as a time series of precipitation with three rainfall events as in Figure 4.28. The optimization variables are the volume of storage and the set point of the pumping system. The objectives are the total cost, the pollution impact and the flooding. Therefore for this experiment we fix the sewer network pipes, using the pre-design. The main objective of this optimization is to minimize the pollution impacts. However, flooding volume is also included to avoid systems that may breach that objective.

Table 4.22 Settings of the optimization algorithm for Run 5

Optimization	Run 5
NSGA II Functions	Default function operators
Population	60
Generations	40
Stoping criteria	Generations exceeded
Runing time (hr)	45.3

Pareto solutions for the storage volume design using three rainfall events
The results of the optimization process are shown in the form of Pareto optimal solutions in Figure 4.32a). The cost of the storage versus the DO sag is shown in Figure 4.32 b). The square point in the figures represents the objective function value for the pre-designed system. As was expected, the solutions that have low storage volume generate more impact in the river. In fact, the pre-designed system is among the worst alternatives in terms of pollution impact generating a DO sag of 6.4 mg/l. That gives a minimum DO in the river of 1.6 mg/l (considering the saturation concentration of oxygen is 8 mg/l). Within the Pareto solution set, there are alternatives that fulfil the minimum standard for the DO in the river. Those that have DO sag less that 4 mg/l imply a minimum DO in the river above the standard (4 mg/l).

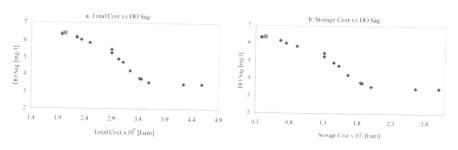

Figure 4.32 Pareto solutions for the storage volume and pump capacity

Table 4.23 shows a comparison of four alternative designs and the pre-designed system. As was observed in the response surface experiment, to maintain the water quality standard in the river, it is necessary to store a significant volume of the runoff. For the pre-designed sewer network, a storage volume the order of 13000 m³ is needed to comply with the minimum DO standard. The solutions tend to leave the pump ratio around the maximum value allowed by the setting of the optimization (RQp=6). That is consistent with the water quality results, because bigger ratios imply that more of the runoff is treated before it is discharged to the river. However, solutions with a higher storage volume (as in alternative 12) tend to have associated with them the lowest pumping ratios to obtain a similar level of impacts on the river quality. The increase in the total cost indicators may appear very significant, but this can be due to the base scenario for the storage computed using the formula A (120 l/hab * Population) being not enough to control the pollution of the receiving system. In conclusion, the experiment demonstrates the benefits of the method proposed by finding solutions that have less impact in the receiving system than the empirical set of rules traditionally used to define the setting of an UWwS.

Table 4.23 Comparison of optimal Pareto solutions for storage and pumping ratio

Design Alternative	Pipe Volume m³	Storage m³	Pump ratio Qp/DWF	Flood x10³ m³	DO min mg/l	Storage cost x10⁶ €	Total Cost x10⁶ €	Cost comparison
Pre-design	5977	2603	6.0	0	1.6	0.45	2.03	-
15	5977	9558	5.7	0.0	2.8	1.32	2.90	43%
16	5977	12697	5.8	0.0	3.8	1.67	3.25	60%
17	5977	14412	5.9	0.0	4.2	1.85	3.43	69%
18	5977	15943	3.7	0.0	4.5	2.01	3.59	77%

Up till now we have solved the problem by letting one of the components of the sewer network be fixed and allowing the algorithm to search on the space of the other variable. The next step is to test the full capacity of the multi-variable and multi-objective optimization algorithm by trying to find optimum solutions for all components (sewer pipes, storage and pump settings) simultaneously.

4.9.4   Optimization of the integrated urban wastewater system

Characteristics of the optimization
For this experiment the integrated model of the system was used. The aim was to use the integrated model to assess simultaneously the risk of flooding in the sewer network and the pollution impacts on the Lili River. The main disturbance of the system was the precipitation time series with one dry period of 6 hours, one precipitation event of 20 minutes with TR=1:20 years followed by 6 hours of dry weather.. The optimization variables are: 25 sewer pipes, the storage volume and the flow ratio of the pumping system. The objective of the optimization is to minimize: the cost of pipe network, the cost of storage, volume of flood and pollution impact (as DO sag). Using the knowledge acquired from the previous experiments, we reduced the search space of the variables to improve the chance of finding optimum solutions within a limited number of function evaluations. Thus, the pipe search space was reduced from 11 to 7 pipe diameters for each pipe in the network; the storage volume was modified from a continuous variable to a discrete variable with 20 steps ranging from 10000 m³ to 15000 m³. The pumping ratio was left as before, in the range 1 to 6 times the dry weather flow. The settings of the optimization algorithm are presented Table 4.22.

Table 4.24 Settings of the optimization algorithm for Run 6

Optimization	Run 6
NSGA II Functions	Default with penalty function
Population	250
Generations	204
Stoping criteria	Average change in the spread of Pareto solutions < 10-6
Runing time (hr)	116.7

Pareto solutions for the sewer network design using integrated approach
The results of the optimization process are shown in the form of the Pareto optimal solutions in Figure 4.33. The cost of the sewer network versus the flooding volume is presented in Figure 4.33 a) and the cost of the storage versus the DO sag is shown in Figure 4.33 b). The square point in the figure represents the objective function value for the pre-designed sewer network. The dots with diamond shape represent the many solutions found with the optimization algorithm. As can be seen in Figure 4.33 a) and b) many of the solutions are sub-optimal (i.e. are optimal for one objective but not for the others). In higher dimensional objective spaces (that is more than two objectives) the number of Pareto solutions increases because one solution may be non-dominated by one of the objectives and therefore maintained in the Pareto set even though it may be dominated by other solutions in another objective space.

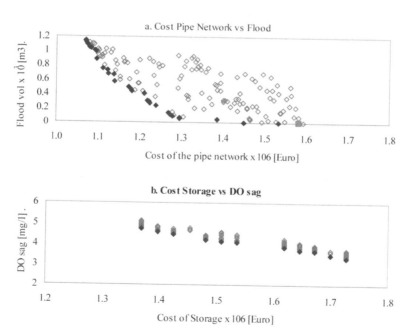

Figure 4.33 Pareto Solutions for the optimization using integrated approach

The Pareto sets presented in Figure 4.33 are the results of running the model two times in series. Each time the model evolved 102 generations and stopped because the average change in the spread was below the threshold defined (<10-6). Due to the high demand of computing time for the optimization, it was decided to assume that the algorithm converged after 204 generations. However, the true Pareto may require more function evaluations. It seems that the NSGAII algorithm has difficulties in exploring areas that correspond to low flooding volumes as seen in Figure 4.33 a) and the storage volume that coincides with DO sag near 4 mg/l in Figure 4.33 b). Zitzler, *et al.* (2001) describe similar difficulties with the NSGAII algorithm when more than two objectives are evaluated. It seems that the algorithm has limitations in the distribution of the solutions and limitations to generate no-dominated solutions vectors that lie in certain areas of the search space (Coello Coello, *et al.* 2001).

From the practical point of view, the many alternatives generated by the MOEA bring a challenge for the decision maker who has now many more options to decide from. This implies a post processing activity that has not been seriously addressed in this thesis. This is a topic for further research. To separate solutions for further analysis, a common sense rule was applied: solutions that produce near zero flooding volume, and cause an impact on the river that does not violate the minimum standard of DO= 4 mg/l are chosen. Table 4.25 shows a comparison of four alternative designs and the pre-designed system. In terms of pipe networks, the solutions found are again the order of 5% to 18% smaller than the pre-design. However, as was expected the solutions are not the same. The reasons can be that the MOEAs are stochastic so for the same problem different solutions can be achieved; another more methodological cause is the limitation in the convergence of the experiments. For the storage volume the selected solutions have two values, which may imply that solution did not fully converge. From these values a storage volume of 10.8 x10^3 m^3 seems to be enough to reduce the pollution impact on the river to levels that comply with the DO standards. The pumping ratio appears to have converged to a value the order of 5.5. That means that during wet weather conditions the treatment plant receives a flow equal to 5.5xDWF.

Table 4.25 Design alternatives using the integrated approach

Design Alternative	Pipe Volume m^3	Storage m^3	Pump ratio Qp/DWF	Flood x10^3 m^3	DO min mg/l	Pipe cost x10^6 €	Storage cost x10^6 €	Total Cost x10^6 €	Total Cost comparison
Pre-design	5977	2603	6.0	0	1.8	1.58	0.45	2.03	-
19	5852	12822	5.5	0.00	4.1	1.59	1.68	3.27	61%
20	5418	12758	5.2	0.07	4.0	1.46	1.67	3.14	54%
21	5362	10754	5.5	0.01	3.8	1.45	1.45	2.90	43%
22	4777	10754	5.5	0.06	4.1	1.30	1.45	2.75	35%

If cost is not an issue, the decision maker may go for alternative 19 in which a bigger volume of pipes and storage may produce amore resilient design for larger precipitation events than the 1:20 year TR selected for the optimization. However, if cost is a driving issue for decision making, the alternative design 22 appears to be the best. The savings are in the pipe network; as explained above it is associated with the reduction of the pipe diameters in the several last pipes of the network (Figure 4.34). With 35% more of the initial cost of the system, it is possible to achieve a near zero flooding volume for TR=1:20 years and a minimum DO in the river above the standard of 4 mg/l. Perhaps this is the main success of the optimization, namely to have achieved an increment in the minimum DO of the river from 1.8 mg/l with the traditional pre-design approach to a 4.1 mg/l by the applying MoDeCo approach. This not only fulfills the standards but it may preserve the life of the ecosystem in the Lili River.

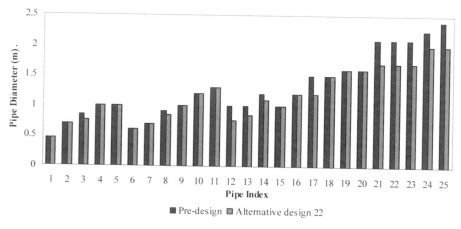

Figure 4.34 Pipe diameters of the sewer network pre-designed and alternative solution 22

The great advantage of these alternative solutions is that they correspond to an integrated analysis of the system in which the synergy between the three components of the system, the sewer, the treatment plant and the Lili River has been included. The solutions are also optimal not only to protect the community from flooding events in the urban catchment but also to protect the environment that receives the discharges from the city.

4.10 Discussion

MoDeCo approach to design a sewer network
In general, the method proposed is capable of finding solutions for the UWwS that fulfil simultaneously both objectives: namely minimize the risk of flooding for the specified rainfall event and at the same time limit the impacts on the river Lili by complying with the minimum concentration for dissolved oxygen.

The design of a sewer network should include the interaction with the other components of an UWwS. The results of this case study demonstrate that an optimum network designed for flood protection, may not comply with the protection of the receiving systems. Fixed rules also prove not to be optimum when the system is evaluated under dynamic conditions. It is demonstrated that customized rules have to be derived for the specific case that is being designed (Vélez, *et al.* 2011).

The design based only on one precipitation event may also not be sufficient to consider critical effects for flooding and pollution. Therefore it is necessary to also include disturbance events that are critical for the objectives of interest. To optimize flooding effects, events characterized by a high volume of precipitation but low frequency (i.e. precipitation with TR=1:20 years) should be used. At the same time, to assess events that may be critical for their pollution impact in the receiving system the events to analyze may be characterized by low volumes of precipitation but more frequently (e.g. precipitation event with TR =2:1 year).

Variations of boundary conditions in the receiving system are a key issue when an integrated view of the design to be addressed. The selection of the flows should be done considering the objectives of interest. For instance, for the Lili River low flows were used to analyze the impacts of pollution (50% permanence flow). However, if flood risk cause by backwater effects in the Lili River is the objective, then the design should include a boundary flow in the river with a return period of 10 or 20 years at least. Thus, for further research, the case study should also address this issue for a more integrated view of the problem.

The generation of knowledge of the behaviour of the system is also a positive result of the implemented method. Here for instance, we learn that with the rational method the system has a level of protection for flooding the order of 20 years return period, even though the initial design rainfall event is 2 years. Therefore, if the objective to be optimized is flooding, the function evaluations must be tested for the precipitation that fulfils the level of protection desired (e.g. 20 or 50 years return period). However, as mentioned by Butler and Davies (2000) the value of this information will be related to the quality of the data used to build the models and the documentation of the model development.

The design of the sewer network is not a hydraulic exercise. Nowadays, regulations are stricter for controlling the pollution impacts caused by drainage from urban areas and effluents of wastewater treatment plants. Therefore, the method proposed here to address this problem gains more importance, since the design of a sewer network has to comply with pollution impacts in receiving systems. A more integrated and holistic approach is the only way this can be done. The MoDeCo approach may assist practitioners in the design of UWwSs that fulfils requirements such as those for sustainable urban drainage systems (SUDS) for flood protection and at the same time for the water framework directive (WFD) for the prevention of pollution impacts in the receiving systems.

Minimization of cost is frequently one of the objectives of the optimization of the sewer network design. The results of the case study show that the optimization procedure can find solutions that are around 15% less costly that the sewer network design using the traditional rational method to size pipes. This cost saving is comparable with the 5% to 30% reduction reported in the literature for optimization of sewer networks (Guo, *et al.* 2008). The reduction of cost has been classified as modest or marginal by other authors but perhaps this should not be considered the main driving force for optimization. In fact, the main benefits of a multi-objective optimization are the knowledge acquired of the system and the possibility to assess a great number of alternatives that will lead to a more informed decision.

Integrated modelling tool
Complexity of the model is case specific and no recipe will work for all scenarios. In the case of Sector 1A, for instance, as the sewer network was the main component to be designed great detail was needed. In contrast the treatment plant model was as simple as possible to capture the effect of variations in effluent flow and composition. Another example, of complexity differences was in the river, where since no backwater effects are included, the kinematic wave was used to represent the variation of flows, but in the sewer network, the fully dynamic approach was used to capture possible flooding events caused by backwater flows. Similarly, the water quality processes were simplified to include only transport in the sewer network to a more complex model in the river that includes advection, dispersion and transformation.

The time of simulation is a critical issue because of the need to assess a great number of alternatives in the optimization. The time is influenced by different factors: the period to be simulated, the simulation time step used in the models, the required output time steps and even the optimization algorithm used and the capacity of the computer. The simulation period is case specific and must be such that the effects of the disturbance on each of the components of the system can be understood. For the case study, the time of concentration of the urban catchment was very short (20 minutes); however, processes of accumulation of dust and dirt in the catchment are of the order of days. The treatment model responds better when left to run some days before the disturbance peak arrives and the effect in the river may be shifted in time as it depends on the discharges from the online storage tank and the treatment plant. Short time steps may be required to guarantee the stability of the solutions of the models. The output time step may also influence the time the model spends on dealing with data. In the case study a 5 minutes time was used, although one minute may describe better the peaks of flow and the water quality components; processing the volume of information generated became a limiting factor for the analysis required.

Perhaps the main limitation to building an integrated model is the information available. The sewer model and the WwTP model are based on the design of the system so there is no data to validate the models. This is not an advantage because most of the parameters needed for the models are based on literature or the expertise of the modeller. This in practise introduces a bias in the modelling system and uncertainty in the results. One of the main limitations of the sewer and treatment plant data is associated with the water quality models. In the sewer model the water quantity estimations, even though are complex to estimate (e.g. infiltration flows), are perhaps better quantified and known than the water quality estimations. Here any modeller of a new sewer network will face a significant gap of knowledge. The build-up and wash-off models used here prove to be a rough estimation of what could happen in the network, but they are highly dependent on the parameters of the function and the dry period before the precipitation event. Here in fact a simplified transport model is used but the transformation process is not an issue as the retention time of the system is very small. More research is needed to assess the impact of water quality models in the sewer to the overall performance of the UWwS.

The calibration and validation of the river model (or receiving system) is important. It is here where the main gap in information can be found. The complexity can be such that the cost of calibrating a model may hinder the whole integrated model. The suggestion from the case study developed here is to select the impact that needs to be addressed and to reduce the boundaries of the analysis to the minimum required to assess the objectives of interest. Significant methodological contributions can be found in the Urban Pollution Manual (FWR 1994); however, case specific adaptations have to be done. For instance, time series for the precipitation events in the sewer and in the river catchment seem to be the best alternative to analyze the impacts in the river. However, this kind of analysis will have a high computing demand limiting the optimization process.

In general the uncertainty may hinder possible benefits for design of UWwS when integrated models are used to analyze alternatives. Therefore, in practical application the uncertainty and the propagation of the uncertainty should be a matter of detailed analysis. Further resources should be adapted to reduce the uncertainty in the models used to produce the function values. A topic for further research is the propagation of the uncertainty and how to present the results of the optimization to the decision makers including the uncertainty.

Optimization algorithm
The design of a sewer network with optimization MOEAs may be limited by the number of variables included in the optimization. The more pipes of the sewer network are included as optimization variables the more difficult it is for the algorithm to converge to a set of that

optimal solutions. Increasing the number of objectives also increases the computing demand and limits the convergence to a Pareto set of optimum solutions. In higher dimensional objective spaces (more than two objectives) the number of non-dominated solutions increases. This presents a much greater challenge, for archiving strategies convergence and good distribution of solutions.

Constraints highly complicate the topology of search spaces, making it more challenging for EAs to reliably identify near optimal or optimal solutions (Coello Coello, et al. 2002). For dendritic pipe networks, the constraint that pipe diameters must be equal or bigger than the up stream pipe may limit the convergence of the MOEA. Within the alternatives of handling constraints, two were implemented: the penalty function and the customization of the NSGAII functions. The default functions of NSGAII are heavily randomized; therefore many of the solutions created are not feasible. This issue was addressed by changing the creation, mutation and crossover functions from NSGA II. However, the results of the algorithm proposed are not significantly better than the traditional penalty function. Thus, further research on that inequality constraint should be done.

The definition of the objective functions is a key factor to finding reliable solutions. Ill-defined problem may not reach optimum solutions or have no solution at all. For the case study, a detailed cost function of the sewer network was developed to improve the reliability of the cost information used to compare design alternatives. The flooding function used, corresponds to the volume that left the nodes and is estimated with the model. Even though it is a good estimator to compare alternatives; it could be improved by including the damage produced in the surface. Pollution impacts were measured with the minimum DO set in the river. An attend was made to use a function that includes the duration that a standard value is exceeded (Schütze, et al. 2002), but the solutions were not as consistent as expected. In Colombia, composite standards with concentration/duration thresholds have not yet been implemented, therefore minimum values are still a good estimate for the case analyzed. However, a composite indicator or the use of more water quality indicators should be explored further.

As with any algorithm that has parameters that need to be tuned, the success of the optimization algorithm in finding a true Pareto Front is highly dependent on a careful selection of parameters and function operators (crossover, mutation). In the case study, a sensitivity analysis of the parameters was performed to find the best settings; however, population size and generation were greatly limited by the computational time. As explained by Nicklow, et al. (2010), the use of evolutionary algorithms requires a carefully designed computational experiment. And that imply having a good knowledge and skill in using the optimization algorithm, and many trial and error tests to understand the behaviour of the specific system being optimized.

The use of known solutions as part of the initial population was a key element to seeding the optimization process. Without the known solutions, the algorithm tends to stall in one or two solutions and does not find the Pareto set. The first seed solution used was the pre-designed system; this was clearly a sub-optimal solution but it helped guide the search. Following in the trial and error test, new, good seed solutions were selected to accelerate the convergence of the optimization process.

One of the limitations found using NSGAII was that the algorithm tends to converge prematurely. It seems that the algorithm stalls in a certain variable space, and the mutation

100

and crossover function do not include the needed variation for the vector to produce new solutions that improve the Pareto set. To deal with this problem, the optimization was run in series, using the final population of a run as the initial population of the following run. This premature convergence and stalling of MOEAs has been described for other algorithms as well (Coello Coello, *et al.* 2001, Zitzler, *et al.* 2001) and may be more critical for high dimensional objective optimizations.

One of the critical issues that require further research is to find a true Pareto Front. Because this case study is an academic exercise, we stop the optimization process having in mind the considering limitations of time, but in a practical application the true Pareto should be found. That may imply a greater number of function evaluations, which is one of the major limitations of the application of MoDeCo. Therefore, further research should consider a method to define when the Pareto set is close enough to an unknown true Pareto set of design solutions. Perhaps the use of hybrid algorithms that use first NSGAII and then a local search algorithm to find the optimal Pareto set will greatly benefit the optimization (Guo, *et al.* 2007).

Perhaps the main limitation of multi-objective optimization in practical application is the long computing time required (Coello Coello, *et al.* 2002, Schütze, *et al.* 2002, Vanrolleghem, *et al.* 2005, Fu, *et al.* 2008). The high computing demand of this research was due to two reasons: 1) the optimization algorithm requires a significant number of evaluations to converge to a near true Pareto set, and 2) the evaluation of the objective functions in the integrated model of the system is computationally demanding. For the case study to find Pareto sets that were acceptable for the analysis it was necessary to evaluate the objective functions up to 50000 times. For the integrated model this requires more than six days of computing. This may seem feasible when it is being done only one time but if you need to run the optimization for many trials, as is normally the case if a proper convergence is to be achieved, and then it becomes a limiting factor of the methodology. Therefore, further research should certainly be focused on reducing the computing time of the optimization process.

Main benefits of MoDeCo approach

- The approach helps to obtain a synthesis of optimum sewer design, storage and pumping flow considering not only the protection again flooding events but also the protection of the receiving system.
- Based on the pollution impacts, flooding protection and costs; the solutions found with the MoDeCo approach have a better performance than the solution designed using the rational method and empirical rules to define the pipe sizes, storage volume and pumping settings.
- The approach enhances the analysis of a great number of alternatives and the Pareto set of solutions gives a variety of design alternatives to be analyzed further by the decision makers.
- The designer and decision makers are better informed of the solutions and their consequences. In fact, perhaps the main benefit of the method is the information generated for each alternative solution.
- The approach is in line with new regulations that enhance the holistic view of the urban wastewater management and the reduction of impacts in the receiving system.

Main limitations of MoDeCo approach
- There is a lack of information to build the integrated model. This is an issue that is not only inherent to MoDeCo; in general, is a limitation for any kind of methodology that tries to address in an integrated way the design of an urban wastewater system.
- The complexity of building an integrated model demands different expertise and skills. In fact, this is not a job that should be done by one person but by a multi-disciplinary team.
- The optimum design depends on the accuracy of a models' prediction. And the uncertainty in the models may threaten the validity of the optimization process. Moreover, the success of the approach relies heavily on the skills, experience and judgement of the engineer who sets up and runs the models.
- The approach has a high computational demand. Two factors influence the computing demand. First, an integrated model of the system is computationally demanding in itself, and second, the optimization process requires a significant number of function evaluations to converge to a set of optimum solutions. This will threaten the use of the method in any practical application.
- The design of an UWwS is a multi-variable and multi-objectives problem. But the more variables and the more objectives that are included, the less probable an optimum solution can be found. Thus a proper level of complexity must be decided on, and a proper experimental design should be prepared.
- As the final results of the process are a set of solutions, they may require further analysis by experts to facilitate the decision making.

4.11 Conclusion

The MoDeCo approach is based on three main concepts: the use of modeling tools, multi-objective optimization algorithms and an integrated view of the UWwS that includes the interactions between sewers, treatment plant and receiving system. In this chapter the approach was successfully implemented in the design of one of the most heavily studied components: the sewer network. A traditional design based on the rational method and empirical rules was used for comparison with the automatic design based on the MoDeCo approach.

The main contributions of this chapter to the research of optimum design of sewer networks are:

- To have expanded the scope of the sewer network design to include water quality impacts in the receiving system.
- The integration of state of the art models with the right level of complexity. In other words, to prove that is possible to avoid the over simplified models that traditionally have been used in optimization of UWwS.
- The problem has been approached as a multi-variable and multi-objective optimization problem. Optimization of sewer networks has traditionally used a maximum of two objectives: cost and flooding damage. Since the aim of this research was to have an integrated design of the sewers, up to four objectives were used in the optimization. The benefits and drawbacks are discussed, thus, contributing to the future applications of multi-objective optimization of sewer networks.
- Carefully described experiments are presented for a non-trivial design problem.

- Methodical approaches are presented to overcome the limitations of data to build an integrated model of the UWwS.
- Two approaches to handling inequality constraints in the design of the sewer network are presented and discussed.

As with any other ongoing research the approach proposed has to be further improved by:
- Developing a more efficient optimization processes. Other MOEAs or a combination of different algorithms (hybridizing) may be used to enhance the possibility of finding global optimum solutions.
- Reducing the computing time by using a parallel infrastructure and parallel algorithms or by using surrogate models to reduce the computational demand of the process based models.
- Quantifying the uncertainty of the integrated model and describing the effects on the uncertainty to the optimized solutions

Perhaps the greatest advantage of the MoDeCo approach for the sewer design is that the alternative solutions corresponding to an integrated analysis of the system in which the synergy between the three components of the system, the sewer, the treatment plant and the Lili River have been included. The solutions are optimum not only to protect the community for flooding events in the urban catchment but also to protect the environment that receives the discharges from the city.

5 Functional Design of Gouda Wastewater Treatment Plant

5.1 Introduction

This chapter describe the application of the model based design and control (MoDeCo) approach for the functional design of a wastewater treatment system. The motivation of this chapter is the need to improve the disturbance rejection in urban wastewater treatment plants. This requires a better process control and as was mentioned by Olsson and Newell (1999) should be the goal for WwTP designers. The disturbances can be categorized as internal; these are the ones produced inside the plant or externally, which are those imposed by the influent flows and wastewater concentrations. The design pumping systems or return streams can cause major hydraulic and nutrient concentration disturbances (Olsson and Newell 1999).

Wastewater treatment processes are more typically moved between operating points heuristically and on the basis of past experience. It has been demonstrated that improvements in transfer time of between two to ten can be achieved using optimal control. Often an optimal control is used to improve the heuristic operational strategies without actually implementing the results automatically on-line (Olsson and Newell 1999). An important development has been made in the optimal design of controllers that are single input and single output (SISO). However, less research has been done on multivariable controllers, multiples inputs and outputs (MIMO).

The problem is that in reality the control of a wastewater treatment plant process is a multivariable problem. It includes multiple operational variables (Internal recycles, air supply, chemical dose, etc) and those variables influence different states variables of the treatment (nitrogen, phosphorous, organic matter, sludge production, etc). Lindberg (1997) use multivariable linear quadratic controllers to optimize the plant performance. In this approach the design includes as inputs: external carbon, internal recirculation rate, and DO set-point and as outputs: ammonium and nitrate in the last aerated zone. Such complex optimization problem has also been addressed using global optimization tools (Moles, *et al.* 2003, Egea, *et al.* 2007, Stare, *et al.* 2007). However, most of these applications are based on benchmark models case studies that do not include the effects of the optimized set point on other processes. In addition the analysis is limited to a one flow condition. For instance, Lindberg (1997) use 9 days of average influent flows and composition and Moles et al (2003) use 7 days with one rainfall event. Even though the problem is a multi-objective optimization problem, most of the authors solve it as a single objective problem by means of giving weights to the objectives, this limits the optimum solution to the one that corresponds to the specific weights, and may limit finding a global optimum solution.

The Gouda wastewater treatment plant has an activated sludge treatment plant. The main concern of Rijnland Water Board (the system operator) is to comply with the future effluent discharge requirements for nutrients total phosphorous (Ptot-P) and total nitrogen (Ntot-N). Therefore, in this chapter the model based design approach (MoDeCo) is implemented to optimize the functional design of the system. The functional design was summarized as the selection of set points for the operational variables: internal recycle (Qir), dissolved oxygen in two sections of the aerated reactor (DO4, DO5) and the dosing of carbon source (VFA). With these manipulated variables, it was possible to reduce the effluent concentration of total nitrogen while keeping the concentration of total phosphorous within set boundaries. The chapter includes a diagnostic of the urban wastewater system, the development of the modelling tools that consist of a simplified sewer model with a plant wide model of the treatment plant. Once the modelling tools are developed they are used within the MoDeCo approach to realize the functional design of the system. At the end of the chapter a demonstration exercise of the anticipatory control concepts is presented.

5.2   Gouda Urban Wastewater System

5.2.1   Sewer system description

According to the Municipal Sewer Plan (Gemeentelijk Riolerings Plan - GRP 2004-2008 in Dutch) the city has 12 drainage areas, which are described in Table 5.1. The type of sewer system is evolving from combined to separate scheme. Thus, it is possible to find five different types of sewer schemes. The system include 175 km of sewer, 200 large and small pump stations and 600 km of stormwater sewers (Cyclus NV 2008). The system has 11 main pumping stations, one per drainage area as indicated in the Table 5.1. The main pumping station is named Bosweg (located in Area No. 4) and has an installed capacity of 3600 m^3/h. Bosweg and Goudseweg stations pump the wastewater to the WwTP. The sewer system also has 34 combined sewer overflows (CSOs) and 11 stormwater overflows that discharge mainly to the open surface waters. Only the CSO of Bosweg discharges directly to the Hollandse Ijssel River which is the final receiving system of the drainage network of Gouda.

Table 5.1. Drainage areas of Gouda sewer system.

No	Name	Combined	Surcharge	Combined	Improved Separate	Separate	Pressure Mains	P.E.	DWF [m³/h]	Area* [ha]	WWF [m³/h]	Pumps station	Capacity [m³/h]	Storage [mm]
1	Bloemendaal					x		26677	352	0	n.a.	Bloemendaal	1080	n.a.
2	J. van de Heijdenstraat	x	x	x	x	x		12968	213	75	507	J. van de Heijdenstraat	720	3.1
3	Achterwillens	x						5589	73	8	227	Achterwillens	300	8.5
4	De Korte Akkeren	x	x	x		x		13423	200	86	257/ 3337	Bosweg	3600	3.1
5	Binnenstad West	x						1641	29	16	131	Nieuwe Haven	150	4.6
6	Binnenstad Oost	x						3106	52	18	198	Tuinstraat	250	2.4
7	F.W. Reizstraat	x	x					11419	156	69	244/ 1262	Reitzstraat	1800	4.0
8a	Goverwelle west				x			3156	40	0	n.a.	Middenmolen laan	126	n.a.
8b	Goverwelle Oost				x			5680	72	0	n.a.	Tempelpolder straat	252	n.a.
10	Goudseweg				x			310	4	0.7	16	Goudseweg	20	7.1
11	Oostpolder in Schielland				x			770	36	9	27	Edinsonstraat	63	3.0

P.E: population equivalents. DWF: dry weather flow. WWF: wet weather flow. *: area drained by the sewer system. n.a:_no apply. Source: adapted with information from GRP Gouda 2004 -2008 (Gemeente Gouda 2004).

Sewer system problems and optimization plan

Gouda has been built on peat and due to this; the ground sinks two centimetres per year. In the past, to reduce the risk of subsidence of the sewers the municipality constructed the system behind houses (in small paths and backyards). However, laying the sewers under private properties hampers their maintenance. The subsidence problem has not stopped, and weakened pipes eventually fail leaking as a consequence and infiltrating groundwater and transporting soil and sand with the wastewater. The sewer scheme in the city centre (Binnenstad West and Binnenstad Oost) is very old and lies under the groundwater level. Infiltration into the sewer pipes leads to a considerable groundwater level change. If the groundwater level in the city drops too low, the wooden foundations under the houses are affected (Gemeente Gouda 2004).

Another problem arises during rainfall events when part of the combined wastewater overflows to open canals and pollutes the surface waters around the city. The water quality of surface water is assessed by comparing water quality data with the standards of the Maximum Admissible Risk (Maximaal Toelaatbaar Risico – MTR in Dutch). This standardisation has been set out by Rijkswaterstaat and applies to all surface water in The Netherlands (Vierde Nota waterhuishouding, 1998). An analysis of the water quality in the surface waters around Binnenstad of Gouda, demonstrated that CSOs have a pollution impact on the canals in the city (Tamboer 2007). Table 5.2 shows the percentile values of water quality measured in 2005. For that monitoring campaign the standards for Phosphorous, Nitrogen, Copper and Dissolved Oxygen (DO) were not compliant with the MRT standards.

Table 5.2. Water quality in binnenstad Gouda compared with MRT standards

Monitoring Point	Phosphorous* 90-perc mg/l	Nitrogen* 90-perc mg/l	Copper 90-perc ug/l	DO 100-perc mg/l
RO148	0.62	2.35	3.6	3
RO434	0.41	4.15	6.0	4
RO581	0.42	3.70	4.6	4
MRT	**0.15**	**2.20**	**3.8**	**5**

*Summer Values from 2005. Source: (Tamboer 2007)

In the Municipal Sewer Plan of Gouda GRP 2004-2008, a plan to address all these problems has been set up, and it includes: stabilize sewer pipes, uncouple the combined sewer system, reduce the infiltration by changing weakened pipes, build sewers in public areas to facilitate access for operation and maintenance and connections to houses. The new sewerage system is planned to be a separate system or an improved separated system. In addition to the separate system, drainage canals will be constructed in such a way that allows the control of the ground water level, via infiltration of rain water or discharge to surface waters to avoid problems with the foundations of the houses. The GRP plan last till 2030, so the replacement and uncoupling of pipes will be done step by step to avoid greater nuisance to the community of Gouda (Gemeente Gouda 2004).

5.2.2 Wastewater treatment plant description

The Wastewater treatment plant (WwTP) of Gouda is designed for removal of organic matter, nitrogen and phosphorus from domestic and industrial waste water of Gouda and two small communities nearby: Gouderak and Stolwijkersluis. The total plant loading design capacity is

140000 Population Equivalents (PE). The DWF estimated is 1375 m³/h and the maximum hydraulic capacity in WWF conditions was estimated as 4350 m³/h.

Figure 5.1 shows the components of the treatment plant and the main flows. The process includes as pre- treatment coarse and fine screens followed by a selector composed of four compartments where grit is removed. After the selector water flows to an anaerobic tank (phosphorous release zone) composed of five compartments. Following these, the treatment process has two ultra-low loaded activated sludge tanks of the RotoFlow type. Each tank has one inner non-aerated compartment (anoxic - denitrification zone) and an outer aerated compartment (aerobic - nitrification zone). After nitrification the wastewater is conducted to four secondary settlers. Phosphorus removal is achieved only using a biological process; though there is the possibility to do a chemical removal of phosphorus, that process is not used. The sludge is thickened and then dewatered in a belt filter press and the solids are incinerated before the final disposal. The treated wastewater is discharged into the Hollandse Ijssel River. Table 5.3 describes the dimensions of the main compartments of the treatment plant based on the design report (Hoogheemraadschap van Rijnland and Tauw Water 1996).

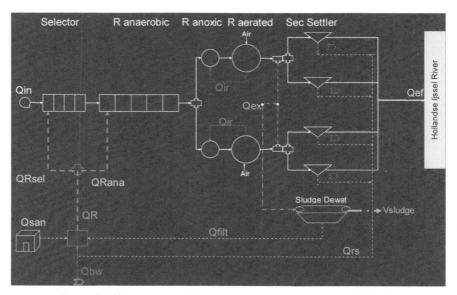

Figure 5.1 Components of the wastewater treatment plant of Gouda.

Note: Qin: influent flows, Qir: internal recycle, Qex: sludge wasted, Qrs: return sludge, Vsludge: volume sludge, Qfilt: filtrated flow, Qsan: internal sanitary flow, Qbw: backwater flows, QR: total return flow, QRsel: return flow to selector, QRana: return flow to anaerobic, Qef: effluent flows. Integer numbers corresponds to average balanced flows in m³/h for year 2004.

Table 5.3 Dimensions of the wastewater treatment plant of Gouda.

Item	Component	Unit/Compart	Volume [m³]	Volume per Tank [m³]	Depth per tank [m]	Area per tank [m²]
1	Screens	2				
2	Bulking Selector	4	688	172	6.5	26
3	Anaerobic Tank (P release)	5	4,812	962	7.0	137
4	Anoxic Compartment (Denitrif)	2	5,028	2,514	6.0	419
5	Aerated Compartment (Nitrif)	2	19,068	9,534	6.0	1589
6	Secundary Settlers	4	7,620	1,905	1.5	1385

Source: (Hoogheemraadschap van Rijnland and Tauw Water 1996)

5.2.3 Functional design of Gouda wastewater treatment plant

The WwTP was fully re-constructed as a new system in 1999, and the change is not only marked in updating the processes but also in the way the system is controlled (operated). The designer used the available state of the art for the control of wastewater treatment plants in 1996. Thus, the system was built with sensors that where fully developed by that time (e.g. dissolved oxygen and suspended sediments). The plant also has a real time control system that is capable of operating actuators in the treatment process. Those actuators were designed with the possibility of operating under variable conditions like variable velocities, flows, water levels or pressures (pumps and blowers), thus giving to the system an operational flexibility and the ability to be controlled using the information collected in the system and the control strategy generated in the controller. This is already an advantage of the Gouda WwTP, because it was conceived including SCADA and RTC components, which is not the case in older designed plants that have no available facilities to install sensors and the devices that can implement controller outputs.

Internal recirculation

The internal recirculation is a flow stream (Qir) from the outer to the inner ring of the activated sludge that is used to bring nitrate-rich water to the pre-denitrification reactor (anoxic tank). The internal recycle can affect the denitrification process in fractions of an hour to some hours by modifying the retention time of the anoxic tank or by disturbing the process with oxygen from the aerated tank. That is why the right flowrate Qir should be selected to guarantee a complete denitrification process. In Gouda WwTP, the recirculation pumps are manually controlled based on temperature conditions and effluent concentrations of nitrate. The recirculation ratio can be varied between 1.5 and 8 times the DWF. According to the design report, the required average daily recirculation ratio for minimum temperatures was estimated as 2.4 times the DWF (Hoogheemraadschap van Rijnland and Tauw Water 1996). In 2010 a new sensor was installed to monitor nitrate and ammonia concentrations at the end of the aerated tank, but to the author's knowledge the information on these sensors has not yet been used to control recirculation flows.

Air supply

The Air flow rate (Qair) is used to add dissolved oxygen to the activated sludge process. The oxygen is fundamental for the aerobic bacteria to degrade the organic mater. But an excess of DO can disturb the denitrification process, influence the organism's growth, the floc formation and the sludge settling properties. From the biological point of view the choice of a proper DO set point is crucial (Olsson and Newell 1999). In the WwTP of Gouda, the air is supplied to the tanks by blowers and distributed through diffusers located at the bottom of the tanks. Diffusers are available in half of the outer ring; there are no diffusers where the recirculation to the inner ring takes place. According to the design report the aeration system has the following capacity:

$$\text{Oxygen transfer capacity} = 2 * 422 \ [kgO_2/h] \qquad \text{Eq. 5.1}$$
$$\text{Blower capacity} = 2 * 1033 \ \text{to} \ 2 * 4135 \ [m^3/h] \qquad \text{Eq. 5.2}$$

Aeration can be controlled from zero when the blowers are off or from 25% to 100% of maximum installed capacity (4135 m^3/h) when the blowers are on. To control the aeration system two strategies were defined:

a. Simultaneous nitrification and denitrification by aerated and non aerated zones
This strategy is to be applied during the day (8:00 – 24:00) and the objective is to maintain aerated and non-aerated zones. The air input was minimized by optimizing the set value of the dissolved oxygen. The control of the number of diffusers in operation is based on a minimum and a maximum airflow per diffuser and as a function of temperature. If the aeration capacity drops below an adjustable minimum value (1033 m^3/h), the aeration system is off. The aeration is switched on again after an adjustable period. The oxygen sensor is placed where recirculation to the inner ring takes place. The values of oxygen measured in the sensors should be close to cero (<0.3 mgO$_2$/l) to avoid disturbance of the denitrification. More over this control should guarantee the simultaneous nitrification and denitrification. To maintain the set point (0.3 mgO$_2$/l) the blowers are locally controlled using a PI controller.

b. intermittent aeration
This control strategy is applied during the night (0:00 – 8:00) and possibly during the winter when low loads of BOD are supply. In intermittent aeration the blowers are off after a set time interval. The aeration is turn on after a set time interval. The un-aerated periods should not exceed 30 minutes. When the aeration is on, the strategy *a.* for simultaneous nitrification and denitrification applied.

Return sludge
The return sludge flowrate (Qrs) is used to control the sludge retention time in the reactor. This variable can influence the system in a time scale of several days or weeks. In Gouda WwTP, the activated sludge is returned from the secondary settlers to the first tank of the selector (see Qrs in Figure 5.1). The capacity of the return sludge pumping station is controlled by frequency adjustments tailored to day / night and DWF/WWF variations. Through the PLC, the return sludge pumping station is linked to the cumulative flow rate of the influent flow meters. Table 5.4 shows the control strategy designed for the return sludge as fractions of the influent flows (Qin).

Table 5.4 Designed control strategy of the return sludge

		Fraction of Qin	Qrs per tank [m^3/h]	Total Qrs [m^3/h]
DWA	Minimum flow		4 * 103	412
	Average flow	1	4 * 344	1375
WWF	Mininum flow	0.418	4 * 455	1818
	Average flow	0.513	4 * 558	2232

Source: (Hoogheemraadschap van Rijnland and Tauw Water 1996)

Excess sludge
The excess sludge flowrate (Qex) is used to control the mass of sludge in the process. The excess sludge and the return sludge are used to control sludge retention time. This manipulated variable will influence the system in a time scale of several days or weeks. In Gouda WwTP, the surplus sludge produced is extracted from the activated sludge tanks. There is the possibility to extract the excess sludge from the pipeline of the return sludge but according to the operators of the plant the system works with the first option mentioned above. To enhance the thickening of the sludge a Polyelectrolyte (PE) is used and then the sludge is loaded in a turbo-drain for mechanical thickening. The maximum size and excess sludge capacity of the turbo-drain system is presented in Table 5.5. The thickened sludge is

then dewatered in a Belt-filter press. Table 5.6 shows the characteristics and capacities of the sludge dewatering system. Part of the filtrated water is re-used to rinse the filter belt but the majority is rejected and sent back to the entrance of the treatment plant to be treated. On average the filtrated water (Qfilt) produced is between 79 – 111 m³/h with a maximum of 146.5 m³/h.

Table 5.5 Maximum size and excess sludge capacities of turbo-drain

Excess Sludge Flow and Capacities	Und	From Activated Sludge Tanks	From Return Sludge Pipe
Maximum sludge production	[kg ds/d]	6230	6230
Reduction in ds 0.1 kg/m3	[kg ds/d]	n.a.	2960
Total sludge production	[kg ds/d]	6230	9190
Sludge content	[kg ds/m³]	4	8
Excess Sludge Flow (Qex)	[m³/h]	109	80
Pe dosage 0.1%	[m³/h]	1.5 - 2.3	2.4 - 3.6
Required hydraulic capacity of turbodrain	[m³/h]	148.8	118.6

Source: (Hoogheemraadschap van Rijnland and Tauw Water 1996)

Table 5.6 Belt-filter press capacity

Excess Sludge Flow and Capacities	Und	Value
Installed capacity	[m³/h]	28
Dewatered sludge dry matter content	[%]	20 - 30
Band rinse water	[m³/h]	20
Filtrate production	[m³/h]	3.3 - 15.2
Amount of sludge	[m³/h]	2.2 - 3.2

Source: (Hoogheemraadschap van Rijnland and Tauw Water 1996)

5.2.4 Performance of the wastewater treatment plant

The main performance indicators of the system are compliant with the effluent standard concentrations. The current standards according to the design report are presented in the Table 5.7 together with the future standards for nitrogen and phosphorous. Based on data from the annual report of Gouda, a comparison of effluent average concentrations versus the standards is presented in Figure 5.2 and Figure 5.3. In terms of the removal of organic mater (measured as COD and BOD) and suspended solids (TSS) the system is very efficient with removal percentages above 90%[i]. With respect to nutrients, removal efficiencies for total phosphorous (Ptot-P) are on average 94%[i] and for Total Nitrogen (Ntot-N) are on average 81%[i], which is low but still fulfills the current standards. However, if we consider the implementation of future standards, with the current historical performance, the system will not be complaint for nitrogen and may have difficulties to be compliant with the phosphorous standards (Figure 5.3).

[i] Removal percentages are base in the annual report developed by Rijnland Water Board.

Table 5.7 Effluent standards for the wastewater treatment plant of Gouda

Parameter		Effluent Standards	
		Current [mg/l]	Future [mg/l]
BOD	Average[1]	< 8	
	Maximum[2]	< 15	
Ntot-N		< 10[3]	< 5[3]
Ptot-P		< 1[1]	< 0.5[1]
TSS	Average[1]	< 12	
	Maximum[2]	< 30	

[1] Average of 10 consecutive measurements, [2] Maximum sample day, [3] Annual average, [4] Only requirement for N-Kjeldahl. Current standard source: (Hoogheemraadschap van Rijnland and Tauw Water 1996), future standards based on personal discussion with Rijnland Water board.

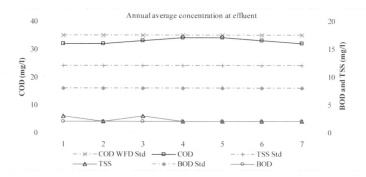

Figure 5.2 Effluent concentration of COD, BOD and TSS versus current standards.

Data source: Rijnland Water Board – Wastewater treatment plant annual report.

Figure 5.4a, shows the time series of data for the Ptot-P_eff at the effluent. The figure also includes a circular mark for the data that is above 0.5 mg/l and matches with a precipitation event. Most of the time, the concentration of Ptot-P_eff is below the future standards. However, there are some peaks above the standard especially in the period June and October of each year. More than half of those peaks coincide with an increment in the effluent flows (circular marks) which in turn is caused by the precipitation events on the urban catchment. Although, there are other sources of disturbance of the treatment efficiency, the data available support the causal link between Ptot-P_eff concentrations and the precipitation events.

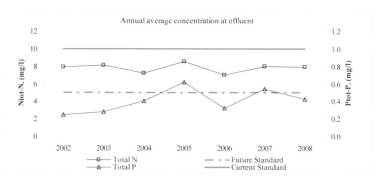

Figure 5.3 Effluent concentration of Ntot-N and Ptot-P versus current and future standards

Data source: Rijnland Water Board – Wastewater treatment plant annual report.

Figure 5.4b, shows the variation in time of TKN-N and Ntot-N in the effluent of the treatment plant. Concentrations of TKN-N_eff that exceed 5 mg/l and coincide with a precipitation event are marked with a circle. 82% of the TKN-N_eff peak values coincide with precipitation events which support the idea that the rainfall impacts the performance of the system negatively. With respect to Ntot-N, the data shows that the system is always above the future standard. Therefore, with the capacity of Gouda UWwS and the current operational strategies the system will not be compliant with future standards. If the data for TKN and Ntot-N are compared, it is possible to conclude that most of the Ntot-N is due to the Nitrate concentrations in the effluent (NO_3-N_eff). This means that in addition to the low efficiencies caused by disturbances, there is a need to improve the denitrification - nitrification process in order to lower the concentration of nitrogen components in the effluent.

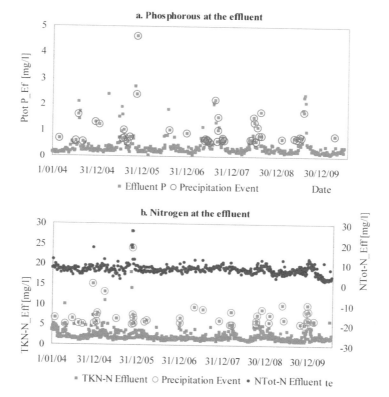

Figure 5.4 Disturbance of nutrient concentration at the effluent. a. Phosphorous b. Nitrogen

In terms of load capacity, the plant works at less than 30% of the capacity it was designed for (140000 PE). And in terms of hydraulic capacity the maximum flow (4350 m³/h) is not even reached during wet seasons. This implies that the existing capacity of the system is appropriated but there is room for improvement in the operation of the system. This coincides with the optimization considerations considered in the document "Options for the Optimization of the UWwS of Gouda" (Langeveld and Jong 2007) in which real time control strategies are mentioned as an alternative.

5.2.5   Formulation of the problem and objectives

Problem statement
The effluent standards for Ntot-N and Ptot-P are going to be stricter in the near future. The data, collected at the effluent of Gouda wastewater treatment plant, shows that on average the system may not be compliant with the future standard for nitrogen and it may have some limitations to be compliant with the standard for phosphorous. Precipitation events over the urban catchment are some of the main disturbances of the system and may hinder the achievement of higher removal efficiencies. Thus, the rejection of disturbances to system needs to be improved together with the nutrient removal efficiencies, all without increasing the operational cost.

Objectives
The main objective of the case study is to improve the performance of Gouda urban wastewater system. The specific objectives are to:
- Improve the rejection of disturbances caused by precipitation events
- Reduce the effluent concentrations of phosphorous and nitrogen
- Minimize the operational cost

Aim of the case study
The aim of this case study is to demonstrate that with the integration of the information generated in the subsystems is possible to design a better operational strategy for the Gouda UWwS.

5.3   Data and Methods

Summary of the research methodology
The general methodology is based on the MoDeCo approach proposed in Chapter 3. In order to address the objectives, first an analysis of the information available and the possible relationships between flows and concentrations was carried out. Then, based on the information available an integrated model of the UWwS was built. The model was used as a tool to understand the disturbances caused by precipitation events. Next, using a forecast of precipitation and the integrated model of the system, a tool to forecast possible disturbances was developed. And finally an optimization process was coupled to find optimum operational strategies for Gouda WwTP.

Data collection and availability
Most of the data corresponds to precipitation on the urban catchment and flows and water quality parameters in the WwTP. A general description of the data available for the development of the case study is presented in Table 5.8. The precipitation data was collected by the Royal Netherlands Meteorological Institute (KNMI). The data corresponds to the Doppler radar precipitation sums every hour on a 1 km grid. The areal average for Gouda was processed using the HydroNET software. The precipitation forecast corresponds to the Ensemble Prediction System (EPS) from the European Centre for Medium-Range Weather Forecast (ECMWF). The data is supply via KNMI to the Water Boards at fixed locations. For Gouda the nearest location is De Bilt, which is about 35km to the north-west.

The data of the WwTP was collected by Rijnland Water Board and corresponds to the monitoring process for control purpose. The data is reliable, because it passed the quality assurance of Rijnland Water Board, but lacks the frequency needed to analyze disturbances in short periods of time (≤ one hour). In the receiving system (River Hollandse Ijssel) there was no information available, only data for the open surface canals that receive part of the drainage of the city.

Table 5.8 Available data of Gouda urban wastewater system

Location		Variable	Metric	Frequency	Period
Drainage Network	one point in the urban catchment	Precipitation	mm	hourly	2005 - 2010
	one point in the urban catchment	10 days EPS forecast of precipitation	mm	6 hours	2010 - now
WwTP	Influent	Flows	m^3/h	hourly	2009 - 2010
		COD, TKN-N, NO_3-N, Ptot-P, TSS	mg/l	5/month	2004 - 2010
		BOD and pH	mg/l, units	2/month	2004 - 2010
	Activated Sludge Procesess	Temp	$^{\circ}C$	2/day	2009 - 2010
		MLSS	kg/m^3	2/month	2004 - 2006
		SVI, Psludge	kg/m^3, ml/g, g/kg	2/month	2004 - 2010
	Effluent	Effluent Flow	m^3/d	10/month	2004 - 2008
		BOD,COD, TKN-N	mg/l	10/month	2004 - 2010
		NH_4-N, NO_3-N, Ntot-N, PO_{4-P}, Ptot-P, TSS	mg/l	5/month	2004 - 2010
Receiving System	Pumps canals to river Hollandse Ijssel	Flows	m^3/s	daily	2000 - 2006
	Sluis canals to river Hollandse Ijssel	Flows	m^3/s	daily	2000 - 2006

The research project lacked resources for a monitoring campaign, so the case study must be considered as a hypothetical case in which some simplifications and assumptions were done as explained below. The analysis of the sewer system was simplified and the focus was put into defining the hydrological response of the catchment to a precipitation event. The greatest detail in the analysis is focused on the wastewater treatment plant. However, important assumptions were needed in order to achieve the degree of detail needed to answer the objectives. Due to the very low impact of the effluents on the Hollandse Ijssel River, this component of the system was very simplified in the analysis.

Data analysis and modelling tools

The data available was analyzed using MS Excel and MatLab. The analysis of the sewer system was centred in the characterization of the dry weather flow (DWF) and wet weather flow (WWF). In addition, a relationship between the wastewater quality variables and information of season, dates, time and flows was explored.

The sewer system was modelled using EPA SWMM. The model is a simplified version of the sewer network in which mainly the rainfall – runoff processes are represented. The schematization and the input data was based on information for the catchment found in the Gemeentelijk Riolerings Plan (Gemeente Gouda 2004) and the analyzed DWF and WWF data. The wastewater quality at the outflow of the sewer network was modelled based on the relationships found between the flows and their concentrations.

The WwTP was modelled using WRc STOAT software. Because of the need to understand the phosphorous and nitrogen process the ASM2d model (Henze, *et al.* 1999) was selected to simulate the biochemical processes. A full model of the treatment plant was developed using the information on the design and the information gathered at Rijnland Water Board for the schematization and inputs. The model was developed following the protocol for dynamic modelling of activated sludge systems as proposed by STOWA (Hulsbeek, *et al.* 2002). However, due to the lack of data, a calibration only under steady state conditions was performed. The calibrated model was tested under dynamic conditions using hourly influent flows and generated influent concentrations. For the integration of the models and the analysis of the results code was developed in Delphi and MatLab.

Assumptions
Sewer

- Assume that the Radar precipitation data is evenly distributed over the urban catchment of Gouda
- The hourly precipitation data are suitable to analyze hydrological processes that occur in the sewer network.
- The EPS precipitation forecast for De Bilt is valid for Gouda. No downscaling techniques were applied

WwTP

- The distribution of the flows was unknown; therefore all lanes of the treatment plant were modelled assuming an even distribution.
- We assume that the model calibrated under steady conditions represents the behaviour of the treatment plant in terms of trend and order of magnitude of the water quality components.

Optimization

- For the purpose of this exercise, we assume that the hydraulic design is optimum. In practice, this is the most reasonable scenario, because modifying the treatment capacity will increase significantly any optimization alternative.

5.4   Integrated Model of Gouda Wastewater System

The integration of the models was achieved by running in sequence the urban drainage and treatment model. The output of the urban drainage system became the input to the treatment plant. Code was developed to create and read the input and output files.

5.5   Model of the drainage network of Gouda

In this section is presented the implementation of the model for the sewer network of Gouda Municipality. The objective of the model is to describe the flows and water quality components of the sewer outflow that is pumped to the treatment plant. Both DWF and WWF conditions are described by the model.

Schematization
The schematization of the sewer network is based on the description of the drainage areas presented above in Table 5.1. The model includes the rainfall - runoff processes in the urban catchment. The storage capacity of the system is included because it modifies the pattern of the outflows. Since CSO outflows are unknown for this research, they are simplified in the model, and when precipitation exceeds the maximum storage the water is extracted from the modelled system as flooding. The modelling time step of 1 hour is defined by the available data of precipitation and flows; (both are 1 hour). Due to the lack of data for the sewer network, the calibration of the model is based on the data collected at the inflow to the wastewater treatment plant. The calibration of the flows and composition of the wastewater is based on data available for the period 01/08/09 to 31/07/10.

5.5.1   Characterization and modelling of sewer outflows

Characterization and modelling of dry weather flows
The analysis of the flows at the influent of the WwTP shows variations in the DWF per season. The different patterns per season are shown in Figure 5.5 a. Flows in winter are on average 20% higher than in the summer season. A similar pattern can be found for the volume of precipitation per season. Thus, lower precipitations (in spring and summer) yield lower average dry weather flows and higher precipitations (in autumn and winter) yield higher hourly flows. Therefore the different patterns of the DWF curves can be the consequence of the infiltration problems with the sewer system. The WwTP can be disturbed by the dilution of the already low loaded influent, so the process control needs to adapt to seasonal conditions. Figure 5.5 b shows the diurnal variation normalized with the 24 hours average DWF (623 $\pm$ 158 m^3/h). This is a typical flow variation of a small population sewer system. Flows are below average between 0:00 to 9:00 hours and above average during the day and up to early night time between 9:00 and 23:00 hours. These hourly variations pose a challenge to the operation of the system and support the control strategies designed for low and high flows mentioned for the current functional design.

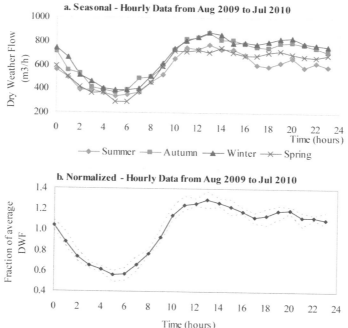

Figure 5.5 Diurnal variation of dry weather flow a. Per Season. b. Normalized with daily average.

Due to the impossibility of acquiring data for the sewer network from Gouda, the DWF was modelled using the diurnal variation curves derived from the available data at the inflow to the treatment plant. The sum of the DWF and infiltration flows creates the modelled DWF. The differences between the measured and estimated DWF are illustrated in Figure 5.6a for a summer period and Figure 5.6b for a winter period. In general, the modelled DWF follows the pattern of the measured values, although it misses the peaks of the data. This high variability of the measured data could be associated with the operation of pumps and the infiltration of nuisance water from the surface canals. This information is unknown for this research so the general pattern described by the DWF model is considered a good enough representation for the purpose of this research. Measurements of goodness of fit support this conclusion. The normalized root mean square error (nrmse) for the summer period, nrmse = 0.7, and for winter period nrmse = 0.9. The coefficient of determination (R^2) for the summer and winter periods are 0.7 and 0.6 respectively. Other indicators are shown in Table 8.9 (Appendix 8.3).

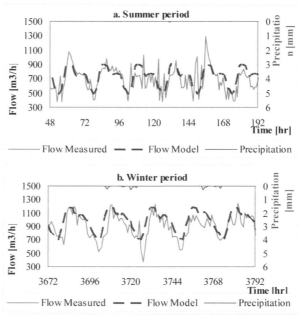

Figure 5.6 Dry weather flow measured and modelled.

Characterization and modelling of wet weather flows
Precipitation events perhaps provide the most significant disturbance of the Gouda sewer system. Since the sewer system is partially combined, precipitation over the urban catchment influences directly the flow variations in the WwTP. The peak WWF measured for the period between August 2009 and July 2010 was 3748 m³/h. Figure 5.7 shows two examples of the variation of the influent flows due to precipitation over the urban catchment of Gouda. Figure 5.7a shows the event with the highest peak precipitation per hour (13.6 mm/h) recorded in the period under analysis. The peak flow is sensed immediately, which means that the concentration time of the catchment is very short (< 1 hour). The recession of the hydrograph is long, which is a consequence of the storage capacity of the sewer system. A precipitation event during the night may overload the system and disturb the efficiencies, especially if the rules for the night time are not adjusted.

Figure 5.7b shows a more frequent type of event with a peak precipitation of 1.2 mm/h. In both events, the peak flow is sensed almost immediately which confirms the small concentration time of the catchment. For short precipitations events, the recession of the hydrograph reaches the DWF conditions faster than was expected. Even though the measured peak flow is below the hydraulic capacity of the treatment plant (4350 m³/h) the impact of precipitation events on the average flow is significant and can be up to 6 times the DWF. The small reaction time available (< 1 h), before flow disturbances are sensed at the inlet of the treatment plant, represents a challenge for the operation of the system.

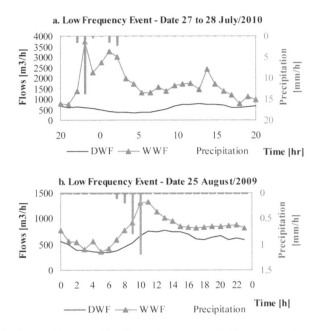

Figure 5.7 Variation of wet weather flows due to precipitation events a. Low frequency precipitation event b. High frequency precipitation event.

Two precipitation events are used to illustrate the model results of WWF. Figure 5.8a shows a high precipitation event while Figure 5.8b shows a high frequency precipitation event. In general, the WWF modelled follows the pattern of the measured values. The time and value of the peaks associated with precipitation events are followed. However, the mechanistic model cannot represent a peak that may have other sources. For instance, the third peak in Figure 5.8a does not correspond to any measured precipitation event. This could be related to missing precipitation data, changes in the pumping schedule or the infiltration of external water to the sewer network. The falling limb of the hydrographs is better approximated in Figure 5.8b, however the recession matches very well with the base flow modelled. Measurements of goodness of fit are presented in Table 8.10 (Appendix 8.3). The indicators show that there is a better performance of the model for the highest precipitation (low frequency). The nrmse and R^2 for the low frequency event are 0.6 and 0.7 respectively; and for the high frequency event they are 0.9 and 0.6 respectively.

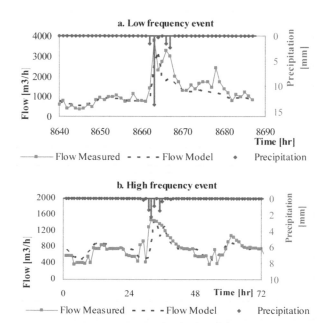

Figure 5.8 Wet weather flows measured and modelled

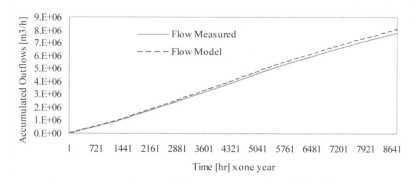

Figure 5.9 One year accumulated flow measured and modelled

The analysis of accumulative flows for one year is presented in Figure 5.9. The year corresponds to the period between 01/08/09 to 31/07/10. The curves show that the model slightly over estimates the flows. The error starts to accumulate after the first five months (3601 hours), which corresponds to the end of autumn and the beginning of winter. For the year analyzed the highest volume of precipitation falls during autumn and winter, thus the accumulation of the error may be influenced by the model response during precipitation events. Even though there is an error in the estimates of the volume, the model represents the general pattern of the hydrographs produced by a precipitation event. Since the interest of this research is to understand the disturbances caused by high precipitation events, the model is considered acceptable for this purpose. Other disturbances caused by the operation of the sewer components or the discharge of nuisance water to the sewer should be considered in further research. A better description of the components of the sewer system and their operation (CSOs and Pumps) and the interaction with the canals may help to improve the performance of the model and to understand the uncertainties. Accepting that the model can represent the behaviour of the sewer network and the general responds of the catchment to a precipitation event, the following steps were to understand and model the wastewater quality components.

5.5.2 Characterization and modelling of the wastewater components

The performance of the WwTP can be affected by the fluctuations of the concentrations of water quality components in the influent. Fluctuations are known to be diurnal and follow the patterns of water use in households. Some of the water quality parameters can also be seasonal, and variations seem to be related more with available organic mater in the sewer catchment during the autumn; the peak of pollution due to intense rainfall events after a dry period can be caused by the re-suspension of sediments. The influent parameters available are temperature, BOD, COD, Ntot-N, TKN, Ptot-P. Since the information of the water quality for the influent is only available on a daily basis, variation patterns were analyzed based on season, precipitation and daily flows. There follows an analysis of the fluctuations per parameter.

Temperature

The water temperature affects the biological process in the WwTP. In sewer systems the temperature is associated with the used water in the household and the ambient temperature. Figure 5.10 shows the variation of temperature for Gouda wastewater using data from 2009 and 2010. The variation of the temperature follow variations of air temperature per month (pointed line). The daily trend of the wastewater temperature can be approximated by a polynomial function (full line).

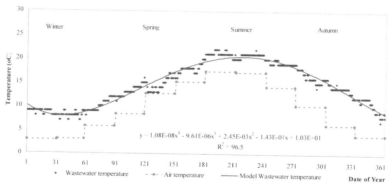

Figure 5.10 Seasonal Variation of Wastewater Temperature

Chemical oxygen demand

One of the main objectives of a treatment plant is to remove the organic matter measured as the chemical oxygen demand (COD). The concentration of COD is highly variable. The variation of COD has been associated with the season, the flows, the re-suspension of sediments in the sewer network, the dirt on the streets, and the use of water by industries discharging to the sewer system, and many other factors. For Gouda the main available information consists of precipitation, averaged daily flows and COD concentrations. An analysis of correlations between those variables is presented in Appendix 8.3 (Table 8.11).

The main correlation of COD is with the flow (Qin), followed by the precipitation one day before (P1), then precipitation two days (P2) and three days before (P3), and finally with the precipitation measured the same day (P0) as the COD.

A scatter plot of flows versus COD is presented in Figure 5.11a. The concentrations of COD decrease with the increment in the flows. This trend may be the result of the dilution of the wastewater in the sewer system by infiltration and rain water. To model the COD different functions were estimated. The relation between influent flow and COD is presented in the Figure 5.11a. The data follow the trend of a power function but with a relatively low coefficient of determination (R2). In order to find a better representation of the data, different classification models were applied. The best model found was the one that uses all the variables of precipitation and inflow (P3, P2, P1, P0 and Qin). The model tree is presented in Appendix 8.3. The comparison of the time series measured and modelled using a power function and the model tree are presented in Figure 5.11b. The model that includes precipitation describes better the series of measured COD. The error of the model tree (nrmse = 0.66) is smaller than the error of the power function (nrmse = 0.79) and it describes better the peaks (Other fitness indicators are presented in Appendix 8.3). For the rest of the experiments the M5 tree model is used to create the COD concentrations at the influent of the treatment plant.

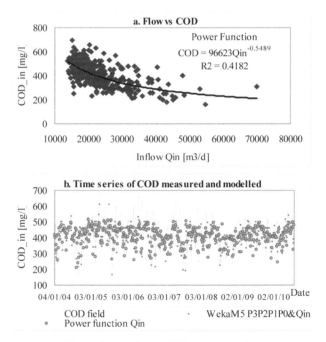

Figure 5.11 Influent COD concentration

Biochemical oxygen demand
The Biochemical Oxygen Demand (BOD_5) is an indicator of the amount of organic mater that is easily degradable by microorganism in the treatment plant. It appears that it is part of what is measured by the COD; a correlation between the two parameters is presented in Figure 5.12a. There is an increasing linear trend of BOD_5 with the increment of COD. It is

122

important that on average BOD$_5$ is 33% of COD; this means that the wastewater has low treatability by biological processes. The correlation analysis that includes precipitation and flows shows that the best correlation is with COD, followed by flows and precipitation one day before the BOD measurement (Table 8.13, Appendix 8.3). A model tree built with these variables (P1, Qin, COD) shows no major improvement in the representation of BOD$_5$. Figure 5.12b shows the time series of BOD measured and the series created with the models. The linear model (nrmse = 0.59) has a very similar fit when compared with the M5 model tree (nrmse = 0.63). Other indicators of goodness of fit are comparable as can be seen in Table 8.14 (Appendix 8.3). Hence for the purpose of the experimental part of this research, BOD$_5$ concentrations in the influent flow are estimated using the linear model as a function of COD.

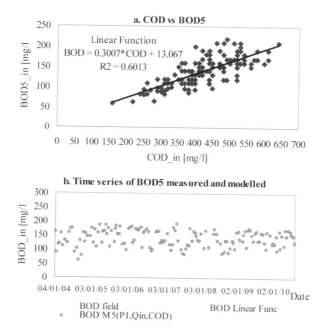

Figure 5.12 BOD concentrations

Nitrogen
The second main objective of a treatment plant is to remove the nutrients from the wastewater. Of main concern are nitrogen and phosphorous. The nitrogen is measured in the influent to the treatment plant as Total Kjeldahl Nitrogen (TKN-N). The correlation analysis that includes precipitation and flows shows that the best correlation is with Qin, followed by precipitation one day before the TKN measurement (Table 8.15, Appendix 8.3). Figure 5.13a shows the variation of TKN-N with Flow. The concentration of nitrogen tends to reduce with the increment in the flow, following the pattern of a power function with an acceptable coefficient of determination (R^2=0.70). Knowing that the precipitation seems to have an effect in the concentration of nitrogen, an M5 model tree was built. The M5 model describes TKN based on the precipitation variables (P3, P2, P1 and P0) and the Flow (Qin). The M5 tree model seems to be better than the power function (Figure 5.13b). According to the indicators of goodness of fit (Table 8.16, Appendix 8.3), the power model produces more error than the M5 model tree (nrmse = 0.6 and nrmse = 0.51 respectively). The power model

tends to underestimate the TKN values. Thus, for the purpose of the experimental part of this research, the TKN concentrations at the influent flow are approximated by the use of the M5 model tree including the precipitation and flow variables.

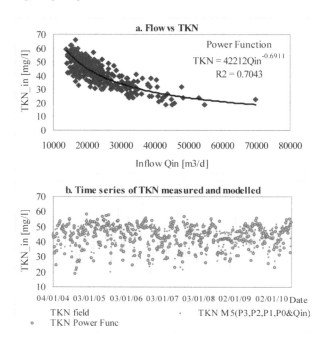

Figure 5.13 Influent Kjeldahl Nitrogen concentration with wastewater flows

Phosphorous
Phosphorous is measured as total phosphorous (Ptot-P) in the influent to the treatment plant. The correlation analysis shows that the best correlation is with Qin, followed by precipitation two days before the Ptot-P measurement (Table 8.17, Appendix 8.3). As for nitrogen, the concentration of phosphorous tends to be reduced with the increment in the flows, following a power function with R^2=0.60 (Figure 5.14a). Following the correlation pattern of the variables an M5 model tree was built for the estimation of Ptot-P. The M5 model describes the Ptot-P based on the precipitation variables (P3, P2, and P0) and the Flow (Qin). However, the M5 model seems not to improve the description of the data better than the power function (Figure 5.13b). According to the indicators (Table 8.18, Appendix 8.3) the power function model produces less errors than the M5 tree model (nrmse = 0.67 and nrmse = 0.77 respectively). Consequently, for the experimental part of this research Ptot-P is approximated using the power function.

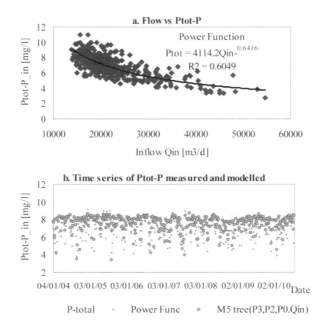

Figure 5.14 Influent Total Phosphorous with the wastewater flows

5.6   Model of the Wastewater Treatment Plant of Gouda

In this section the implementation of the model for the wastewater treatment plant of Gouda is presented. The first objective of the model is to understand the effect of disturbances on the process. Special attention is given to modelling the removal processes of the phosphorous and nitrogen components. The second objective is the optimization of the control strategies; which implies a more detailed description of the control of the process. Both, the disturbances and the control require modelling time steps that allow a description of the processes ($\leq$ 1 hour).

Schematization
All the treatment processes of the full scale treatment plant were included in the model. The biological processes were represented using the ASM2d model implemented in STOAT. The model includes the selector, anaerobic, anoxic and aerated compartments. The settlers were modelled using the implementation of Tackas model in STOAT, including the biological processes described in ASM2d. A simplified version of the thickeners for the sludge handling was used. The distribution of the flows was unknown; therefore all lanes of the treatment plant were modelled assuming an even distribution of flows. The dimensions of the components correspond to those presented in Table 5.3 from the design of the system. Figure 5.15 shows the schematization of the model in STOAT.

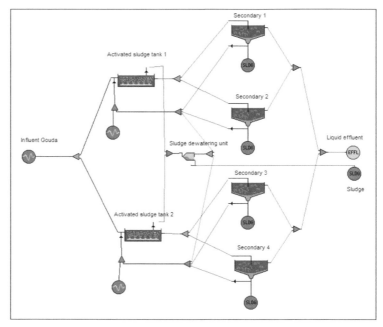

Figure 5.15 Scheme of the wastewater treatment plant model in STOAT

Model structure

The model structure for the hydraulics of the WwTP is based on the specifications of the design and the current operation of the treatment plant. The activated sludge tanks in Figure 5.15 have seven compartments: one for each selector, anaerobic and anoxic compartments and four for the aerated compartment. The compartmentalization of the aerated zone was calculated using the formula presented in the manual of STOAT and the total volume was evenly distributed in the four compartments. The settler was divided in 10 vertical compartments with the same height following the recommendations of the STOWA protocol. The operation of the aerators was modelled using PI controllers and fixed set points for each stage of the activated sludge tanks. The return sludge (Qrs) was set up as a ratio of the influent flows (Qrs = 1*Qin). The return sludge flow was evenly collected from the lower compartment of the secondary settlers and connected to the inlet of the activated sludge tanks. According with the personal communication with the operators of the treatment plant this is the more frequent operation of the system. The wasted sludge (Qex) is taken from the seven compartment of the activated sludge tank. A continuous excess flow rate was specified. The internal recycle (Qir) was set up as a ratio of the influent DWF (Qir = 2.4DWF). The recycle flow is taken from the seven compartment (last aerated zone) and return to the third compartment (anoxic zone).

5.6.1 Characterisation of flows and water quality components

The flows were characterized using the historical data measured at the treatment plant. Data from the year 2004 was used to estimate daily average influent, effluent and internal flows and compositions. The data were checked using mass balances for the flows, COD, dry matter, nitrogen and phosphorous. following a similar approach as proposed by (Meijer, *et al.*

2002). Table 5.9 shows the balanced flows for the treatment plant. The mass balances established to check the data are presented in the Appendix 8.3, Table 8.19.

Table 5.9 Balanced flows of the wastewater treatment plant of Gouda for 2004

Flow Description [m³/d]	ID	Average	Type	Balanced	Comment
Influent Flow	Qin	23237	Measured	23237	Meausured
Effluent Flow	Qef	20573	Calculated	25660	Calculated: Qin+Qsan+Qfil-Qex
Excess Sludge from Activated Sludge Tank	Qex	2664	From Scheme	929	Balanced by Ptot-P
Retourn Sludge Flow	Qrs	33000	From Scheme	23237	Calculated: 1*Qin
Secondary Settler Backwaters	Qbw		Unknown	0	Assumed
Efluent Filtrated from Dewatering	Qfil	2688	From Scheme	952	Calculated: 1.025*Qex
Sanitary and Lab and other internal flows	Qsan	2400	From Scheme	2400	From scheme
Flow Retourned	Qr	38088	Calculated	26589	Calculated: Qrs+Qbw+Qfil+Qsan
Internal Recirculation flow	Qir	72000	Design: 2.4*DWF	55768	Calculated: 2.4*Qin

From Scheme: are values measure by Rijnland Water Board in 1998 and given as a scheme in a personal communication.

The average concentrations of the wastewater components measured in 2004 are shown in Table 5.10. The data include the concentration at the influent and effluent of the treatment plant and some operational parameters measured in the activated sludge. The fractionation of the wastewater components was done using the STOWA method (Roeleveld and van_Loosdrecht 2002) and the STOAT manual. The detailed fractionation of the wastewater components for the average concentrations of 2004 is presented in the Appendix 8.3 Table 8.20.

Table 5.10 Average concentration of wastewater components for 2004

Parameter	unit	Influent		Effluent		Activated Sludge	
		Average	± Stdev	Average	± Stdev	Average	± Stdev
Flow	m³/d	23237	8931				
BOD	mg/l	153.6	47.8	2.4	1.6		
COD	mg/l	371.5	86.7	32.8	8.1		
TKN-N	mg/l	39.2	9.4	2.8	1.9		
NH4-N	mg/l			1.3	1.3		
NO3-N	mg/l	0.4	0.4	4.5	1.4		
Ntot-N	mg/l	39.6	9.5	7.2	2.0		
PO4-P	mg/l			0.3	0.4		
Ptot-P	mg/l	6.1	1.3	0.4	0.4		
TSS	mg/l	99.2	44.1	4.1	3.3		
Temp	°C					15.2	3.7
MLSS	kg/m³					4.4	0.5
SVI	mg/l					76.5	6.8
P(sludge)	g/kgSS					32.5	2.4

5.6.2 Calibration of steady conditions

The balanced flows and the influent concentrations fractionated were used to simulate a steady state condition of the treatment plant. The model was fed with constant values for 200 days to guarantee that a stable condition was reached. Initially the default parameters of STOAT for the stiochiometric and kinetic parameters and the switching coefficients were used. Then, modelled effluent concentrations were compared with the measured data. To establish an initial calibration for the steady conditions, the parameters were adjusted to the values proposed in the description of the ASM2d model (Henze, et al. 1999). The 80 parameters and coefficients required for the STOAT model and the values used for Gouda are presented in the Table 8.21 (Appendix 8.3). Table 5.11 shows the parameters that were modified to reduce the differences between the measured and modelled effluent concentrations for the steady state condition.

Table 5.11 Parameters modified to calibrate the WwTP model of Gouda

Item	Parameter	ID	STOAT Default	Value used for Gouda
2	Fractional hydrolysis rate, anaerobic conditions (-)	ηfe	0.1	0.4
48	Half-rate constant temperature coefficient (1/°C)	a	-0.10986	0.10986
70	SCOD half-rate constant (fermentation) (mg COD/l)	Kfe	20	4
22	P content of inert particulate COD (mg P/mg COD)	iP_{XI}	0.01	0.025
	Ratio Particulated COD Degradable and inert (fXs = Xs/(Xs+Xi))	fXs	0.51	0.43
16	N content of inert soluble COD (mg N/mg COD)	iN_{SI}	0.01	0.02
64	NO3 half-rate constant (heterotrophs) (mgN/l)	KNO3	0.5	0.2
68	NH4 half-rate coefficient (autotrophs) (mg N/l)	KNH4	1	0.8

i. Calibration of the sludge composition and production

First the sludge composition was fitted to the balanced total phosphorous in the influent, effluent and sludge. Ptot-P at the effluent was calibrated by modifying the P-content of the inert particulate COD (iP_{XI}). The value of iP_{XI} was increased, not only in the stiochiometric parameters but also in the influent characterization. To fit the model to the particulate COD balance, the influent ratio of particulate degradable COD and the particulate inert matter (fXs) was decreased, so the influent characterization of COD was also modified.

The aeration in the seven stages of the activated sludge tank was set as follows: for the first three tanks no oxygen was supply and for the last four tanks the aeration was controlled using the PI control with different set points. The set points were: 1 mgO2/l for the fourth tank, 0.5 mg/l for the fifth, 0.3 mg/l for the sixth and 0.1 mg/l for the last one.

Initially the parameters of the settlers were estimated using the WRc correlation with Sludge Volume Index (SVI@3.5) as described in the STOAT Manual. The value of the SVI was estimated as an average of the selected MLSS that gave approximately 3.5kgSS/m³. To improve the effluent concentration of the TSS, the settling velocities of the secondary settler were increased by reducing the SVI value from 76 to 70 ml/g. After these modifications, a good fit between the measured and the modelled sludge characteristic was achieved. Table 5.13 shows the results of the calibration for the sludge composition and production.

Table 5.12 Sludge composition and production measured and modelled

Parameter	unit	Average 2004	Model
Sludge Production	kgSS/d	4063.0	4023.9
Sludge Composition - MLSS	kgSS/m³	4.4	4.4
P content in sludge	gP/kgSS	32.5	30.8
Sludge age - SRT	d	31.1	30.8

ii. Calibration of nitrogen components

To fit the effluent concentration for NH_4-N, NO_3-N and Ntot-N, three parameters were modified. The N content of inert soluble COD (iN_{SI}) was increased to improve the fit of Ntot-N in the effluent. The NH_4-N and NO_3-N were adjusted by modifying the NO_3 half-rate constant (KNO_3) for heterotrophs and the NH_4 half-rate coefficient (KNH_4) for autotrophs.

A comparison of the average effluent concentrations for the 2004 data and the model effluent is presented in Table 5.13. The calibration for steady conditions shows a good fit for the effluent characteristics. Although there are differences, the values obtain with the model are within the standard deviation of the measured values.

Table 5.13 Average measured concentrations in 2004 and modelled effluent concentrations

Parameter	unit	Average Effluent	Model Effluent	Difference
Flow	m³/h	1068.0	1067	0%
TSS	mg/l	4.13	5.74	-39%
COD	mg/l	32.84	33.35	-2%
TKN-N	mg/l	2.81	2.59	8%
NH4-N	mg/l	1.27	1.29	-1%
NO3-N	mg/l	4.49	4.87	-8%
Ntot-N	mg/l	7.23	6.82	6%
PO4-P	mg/l	0.26	0.34	-32%
Ptot-P	mg/l	0.39	0.38	4%

5.6.3 Verification for dynamic conditions

Once the model was calibrated under steady state conditions, its performance was tested under dynamic conditions. For this purpose, time series of data at the influent, effluent and within the process are required. In addition, data of variations in the operational strategies have to be included in the model. In order to test the model for Gouda, information for the year 2009 – 2010 was used. The main information available is the hourly influent flows of the treatment plant and about 60 daily average concentrations of the main parameters (COD, BOD, TKN-N, Ptot-P, TSS). Therefore, the time series for the water quality components was generated using the correlations shown above (Figure 5.10 to Figure 5.14). The daily concentrations estimated with the correlations were assumed constant during 24 hours to build the time series for the influent flows. The calibration parameters and the operational parameters were left as in the steady state calibration. The initial conditions of the reactors were assumed to be those found under the steady state conditions.

In order to assess the performance of the model under dynamic conditions, a comparison was made of the effluent concentrations modelled and measured. The data available corresponds to approximately 120 data of average concentrations measured at the outflow of the treatment plant. To be able to compare the results with the measured data, the hourly values of the model were aggregated into daily average concentrations. Figure 5.16 to Figure 5.19 show the results of the model and the average effluent concentrations for COD, Ntot-N, TKN, NO3-N and Ptot-P.

The model replicates the general trend and the magnitude for COD and Ptot-P. However, the model results are slightly higher than the measurements. For the nitrogen components there is a good match with the trend and with the magnitude of the variables for the first six month. However, from January 2010 there is a mismatch in the trend and in the magnitude of the variables. The main difference is in the NO_3-N which affects the other two components: TKN and Total Nitrogen. The historical effluent records (Figure 5.3) show that on average the Total Nitrogen was around 8 mg/l, and the last part of the year 2010 shows values below 4 mg/l. This may be the consequence of changes in the operation of the system that were unknown to the author.

Even though the fit of the model against the measured values could be better, for the purpose of this case study we assume that the model represents the general trend and magnitudes of the effluent concentrations of interest.

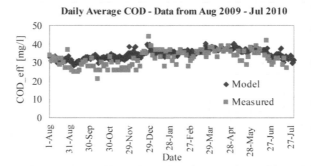

Figure 5.16 Effluent Chemical Oxygen Demand measured and modelled

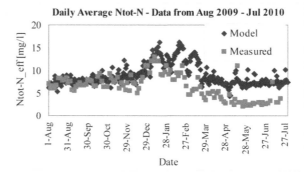

Figure 5.17 Effluent Total Nitrogen measured and modelled

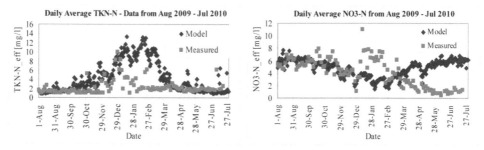

Figure 5.18 Effluent Kjeldahl Nitrogen and Nitrate measured and modelled

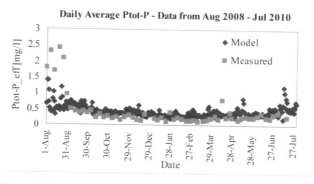

Figure 5.19 Effluent Total Phosphorous measured and modelled

5.7   Functional Design of Gouda WwTP Using MoDeCo Approach

In a WwTP with highly variable load there is strong need for a continuous adjustment, not only of the individual control loops, but the complete plant operation in order to ensure a stable biological nutrient removal. Once the modelling platform for the case study is in place, we can implement MoDeCo approach to optimize the operational settings of the WwTP in order to improve the removal efficiencies of phosphorous and nitrogen. The WwTP use different sub-processes to achieve the biological removal of N and P. Sub-processes in a simple sequence can be affected by what is done in previous sub-processes. Recycle streams increase the interactions between them, and changes in the recycle can be reduced or exaggerated after they have been transmitted around the circle. Because of the interaction between the processes, we need a plant wide evaluation of the operational strategies to achieve the overall goals. But before we optimize the control strategies, we need to understand the limiting factors of the N and P removal processes.

5.7.1   Limiting factors of the N and P removal processes

The biological removal of phosphorous and nitrogen are realized in the anaerobic, anoxic and aerated reactors. Some of the factors and the variables that may influence those limiting factors are summarized as follows.

Phosphorous removal (anaerobic reactor):
- Lack of readily biodegradable organic matter (VFA) is one of the main limiting factors. This could be improved by fermenting sludge in the selector at the beginning of the treatment plant or adding a carbon source.
- During wet weather, the formation of VFA may be limited by oxygenated water entering the system (anaerobic tank). This effect may be avoided by controlling the runoff combined with the wastewater that is pumped to the treatment plant (Qin).
- During a warm season, slow growing methanogens may be favoured. So a sufficient but short retention time is needed to guarantee that methanogenesis is not taking place.

Nitrogen removal (anoxic and aerobic reactor):
- Oxygen level in the anoxic reactor has to be kept at minimum. The DO concentration in last part of the aerobic reactor has to be close to zero otherwise the internal recycle (Qir) will bring oxygen to the anoxic zone.

- Reaction rate can be controlled in fractions of hours by manipulating the influent DO concentration and carbon source (VFA).
- The retention time of the anoxic zone can be changed in periods of fractions of an hour to hours by manipulating the influent flow rate (Qin) and the nitrate recirculation (Qir).
- Reaction can be controlled using the profile of nitrate at that outlet of the anoxic zone. This implies that new sensors have to be placed in the treatment plant.

5.7.2 Definition of the system

Following the objectives of the case study, the boundaries for the optimization of the functional design are closed around the three main reactors involved in nutrient removal: anaerobic, anoxic and aerated (Figure 5.20). Although, the focus is on the optimal operation of those reactors, a plant-wide view is still preserved and some broader aspects like the interaction with the sewer system will be introduced below.

The processes in a wastewater treatment plant have different dynamics, from slow time scales (days to weeks) for biomass growth to fast (minutes to hours) for flow dynamics. The time scale of interest for the optimization of nutrient removal is the order of hours to a few days (medium time scale).

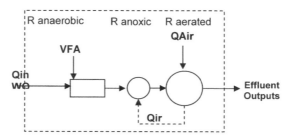

Figure 5.20 Scheme of the boundaries for the functional design

5.7.3 Definition of the inputs and outputs of the system

There are two types of inputs: those that we can manipulate in the system (degrees of freedom) and those that we cannot (disturbances). The degrees of freedom of the system are defined as the set points of the internal recycle (Qir), the air flow rate (Qair) considered by the DO set point of the aerated reactors and the dose of ready biodegradable organic mater (VFA). The disturbances of the system include the inflow (Qin) and wastewater composition (WQ). The main outputs of interest are the state variables of the system for Nitrogen and Phosphorous. However, other state variables like the effluent COD and TSS or sludge production and composition, must be considered in order to preserve the integrity of the treatment.

5.7.4 Definition of the objective function and constraints

The objective function for the optimization includes two types of measures: environmental impacts and use of resources. The environmental impacts can be measured as the effluent concentrations of nitrogen and phosphorous. The objectives related with the resources can be measured as the energy cost for air supplied and the pumping of internal recycles, and the chemical supply when used. The functional design of the WwTP can be posed as a multi-objective optimization in which the aim is to find the combination of set points for Qir, DO and VFA that minimizes the three objectives Ntot-N and Ptot-P at the effluent (f_{TN-N} and f_{TP-P}) and the operational cost (f_{TCost}). Mathematically the problem can be stated as follows:

$$MinF(x) = \{f_{TN-N}(x), f_{TP-P}(x), f_{TCost}(x)\}$$
$$where \quad x = [Qir, DO]$$
$$and \quad x_{iL} \leq x_i \leq x_{iU} \quad x \in \Re$$

Eq. 5.3

$$f_{TN-N} = \overline{(Ntot - N)}_{Ef}$$
$$f_{TP-P} = \overline{(Ptot - P)}_{Ef}$$

Eq. 5.4

$$f_{TCost} = \sum_{i=1}^{AirPump} Cu_i V_i + CqQ_{ir}t$$

Equation 5.1. Multi-objective function for the functional design of the treatment plant

The variables (x) are constrained by the installed capacity of the equipment. The internal recycle is constrained by the capacity of the pumps that according to the design is 1.5*DWF < Qir < 8*DWF. The air flow rate is constrained by the location of the diffusers and their capacity. The diffusers are in the first 2/4 of the aerated tank, so that means that we have two DO set points to fix and the range was estimated as 0 < DO < 5. There is no equipment in place for the dosing of additional carbon, so there is no equipment constraint for VFA.

5.7.5   Response surfaces of internal recycle and dissolved oxygen

The functional design of a WwTP can be seen as a complex multi-objective optimization. Knowing that the variables are continuous makes it computationally very expensive to do an exhaustive search of all the possible combinations. However, it is possible to assume that the set points are discrete variables and to map the response surfaces of the objectives. To do the analysis a wet period of 10 days with hourly input data was used. The initial conditions in the reactors were left as from the end of the steady state conditions. Considering the current status of the Gouda WwTP, only the three variables that can be manipulated were considered in the search: DO set points in the first and second part of the aerated reactor and the internal recycle (DO4, DO5 and Qir). The operating space of these variables was partitioned in 25 steps. Thus, for an exhaustive search, the possible combination of set points is 25^3 which mean that the model of the system run 15625 times for an elapsed time of around 28 hours with a processor Intel core at 3 GHz.

Figure 5.21 shows scatter plots of the objective functions. For the same operational cost there is a wide range of possible Ntot-N effluent concentrations (Figure 5.21a), which demonstrates the need for a proper selection of the set points. Even though the Ptot-P concentration is less sensitive, it is still important to avoid nuisance in the removal process of phosphorous (Figure 5.21b). The optimal set points should bring a trade-off between the three objectives, and that solution most probably will lie within the red lines drawn in Figure 5.21.

Even with the coarse grid of operating points it is very complex to map the response surface of three variables with three objective functions. Therefore, in Figure 5.22 only a sample of the response surfaces is presented. The first group of figures named A1 to A3 represent the sensitivity of the objective functions to the DO set points. Nitrogen and operational cost are the most sensitive, which was expected because the oxygen dissolved affects directly the

133

nitrification reaction (Figure 5.22 A1). In addition the relation with cost holds due to the fact that higher DO set points imply higher air pumping rates (Figure 5.22 A3). In contrast the phosphorous (Figure 5.22 A2) behaviour is insensitive to both set points DO4 and DO5. However, the high peak at the corner of the low DO set points shows that the settings made disturb the removal of biological phosphorous.

The right sides of figures B1 to B3 show the sensitivity of the objectives to the DO4 and Qir set points. The internal recycle is very influential on the nitrogen and operational costs. This effect was expected, since the main purpose of Qir is to bring nitrate into the denitrification process. However, the phosphorous sensitivity was not expected. There is a trend towards getting higher Ptot-P concentrations with the incremental removal of Ntot-N. This demonstrates the important interaction between the processes in the treatment plant. It is important to understand that the relationships found varied with the variation of the external inputs (influent flows and compositions). In conclusion, the response surfaces show the effects of the selected variables on the objectives and support the evaluation of the operational set points as a multi-variable and multi-objective optimization. Thus, the next step is search for optimum set points with the support of optimization algorithms.

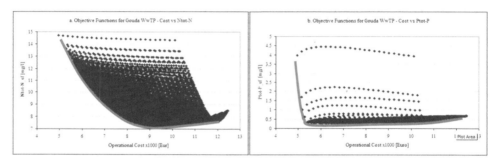

Figure 5.21 Scatter plot of objective functions

5.7.6 Optimization of WwTP set points

The final stage in the functional design using MoDeCo approach is to optimize the set points. As it was presented in the Section 5.2.3, the current functional design is mostly defined for DWF and WWF and for low temperature conditions. In order to find the optimum set points four scenarios that combine influent conditions and temperatures were analyzed. For each scenario 10 days hourly data of the influent flows and composition were created using the sewer model. To build up the DWF scenarios the diurnal DWF curves for winter and summer (Figure 5.5) were used. For the WWF scenarios, the time series was composed of four precipitation events with different return periods (1:1, 1:2, 1:3 y 1:6). The duration of each event is one hour, and they are separated by 48 hours to avoid overlapping effects. The selection of the return period and the duration was limited by the available precipitation data; however, the event can be regarded as representing average wet weather conditions. The influent composition was estimated using the relationships shown in Section 5.2.3; the daily average concentrations were assumed constant during the day but the load varied with the hourly variation of the flows. The time series used in each scenario are shown in the Appendix 8.3. The initial conditions of the reactors were assumed as those found at the end of the steady state simulation, and can be regarded as average initial conditions.

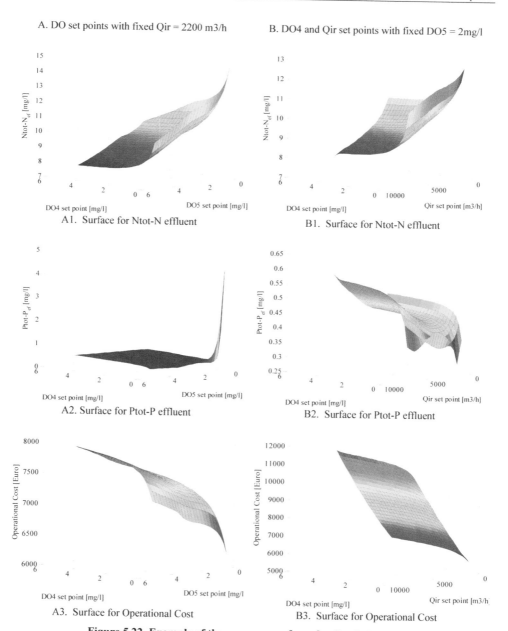

A. DO set points with fixed Qir = 2200 m3/h

B. DO4 and Qir set points with fixed DO5 = 2mg/l

A1. Surface for Ntot-N effluent

B1. Surface for Ntot-N effluent

A2. Surface for Ptot-P effluent

B2. Surface for Ptot-P effluent

A3. Surface for Operational Cost

B3. Surface for Operational Cost

Figure 5.22 Example of the response surfaces for Gouda WwTP

The objective function to be minimized corresponds to the one presented in Equation 5.1. Thus, three variables (DO4, DO5 and Qir) were used to optimize three objectives (Ntot-N, Ptot-P and Operational Cost). The set points for the return sludge and excess sludge were left as in the original design of the WwTP because the main objective is to improve the removal efficiency of nutrients. However, the effect of the sludge production and composition was assessed in the optimized solutions in order to check that the integrity of the process is maintained. In addition, two soft constraints were included by the use of penalty operational costs (25000 Euros) for solutions which average Ntot-N > 14 mg/l or Ptot-P > 2 mg/l. This constraints force the optimization algorithm to fine solutions that are of interest for the design. The optimization algorithm evaluates 2000 combinations of set points in an average time of 4 hours in a processor Intel core at 3 GHz.

In order to compare the results of the optimization, a base scenario was setup using the set points of oxygen found for the steady calibration and the internal recycle ratio proposed in the functional design of Gouda WwTP. This is, the vector of set points [DO4= 1 mg/l, DO5=0.5 mg/l, Qir= 2.4*Qin m^3/h] was used to estimate the objective functions in each scenario and to compare with the solutions of the optimization.

The results of the optimizations are presented in the form of Pareto frontiers in Figure 5.23. Each figure corresponds to a scenario of dry and wet weather flows combined with winter and summer conditions (low and high temperatures). For each scenario two Pareto fronts are plotted, the face of Cost versus Ntot-N_ef in diamond dots and the face of Cost versus Ptot-P_ef in square dots. In addition, there are two dots that represent the objective function values for the base scenario, the triangular dot for nitrogen and circular dot for phosphorous. The rectangular shape demarks the values of the objective function for selected set points.

Figure 5.23 shows that per objective the optimization algorithm may find solutions with better performance than the ones found with the set points used as base scenario. However, the selection of a better solution for nitrogen implies the deterioration of the performance of phosphorous removal. Since the manipulated variables should influent mainly the nitrogen removal processes, the contradictory behaviour can be attributed to disturbances generate by the set points in the processes of phosphorous removal. The set points of DO and Qir affect the reactions in the anoxic and aerated reactors. Thus, the optimized set points may tend to favour the growth of heterotrophs organisms and the denitrification process. By doing these, the set points may limit the growth of PAO organisms and the P uptake process. As mentioned by Olsson and Newell (1999) in a low loaded treatment plant the heterotrophs will compete for VFA with the PAO organisms, affecting the relation between P uptake and P release.

In order to select the optimum set points for each scenario, a heuristic approach was carried out. The selection rule is as follows:
"The optimum combination of set points is the one that produces the smallest concentration of Ntot-N, and at the same time generates a Ptot-P concentration of about 1 mg/l and generates the less operational costs".

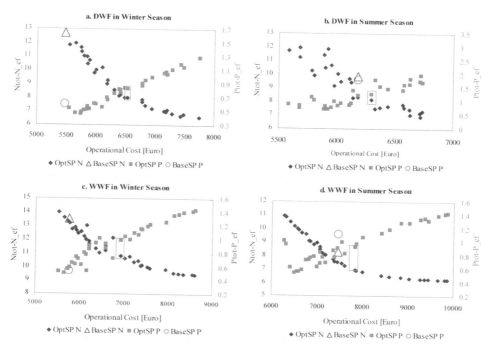

Figure 5.23 Pareto front of the optimization of set point for Gouda WwTP

The rule implies a weighting factor that prioritizes the removal of nutrients over cost, but in practical applications this may be the other way around. Nevertheless, the selection rule pursues the need to comply with future water quality standards. Following this rule, the Pareto solutions were classified and the best was selected for each scenario (the rectangular shape in Figure 5.23 a to d). The selected set points are presented in Table 5.14 together with the values of the objective functions. As can be seen, with respect to the removal of nitrogen, the selected optimum set points performed better than the base operational set points. On average, the total nitrogen concentration is 23% lower in the optimized scenarios. For phosphorous, the effluent concentration for the selected optimum set points is higher for the winter conditions and slightly lower for summer. The operational cost increases 12% on average for the four optimized scenarios. However, this could be turned economically beneficial, if the cost of pollution in the effluent discharge is included. For instance, including the payments of levies for discharged oxygen demanding substances will probably turn the balance in favour of the optimized set points.

When the optimum set points are compared with those from the base line, two trends can be distinguished. The first is that the internal recycle is on average bigger for the optimized scenarios. This trend can be attributed to the fact that higher Qir rates bring more nitrates back to the anoxic zone to be denitrified, and as a consequence there is less concentration of total nitrogen in the effluent. The second trend is that DO5 set points are higher than DO4. These results were somehow unexpected because we may tend to expect decreasing DO in the set points from the entrance of the aerated reactor (DO4) towards to the exit (DO5) following the load of organic mater and nutrients.

Table 5.14 Summary of selected set points and its performance indicators

Operational Scenario	Qrs	Qir	DO4	DO5	Ntot-N_ef	Ptot-P_ef	Cost
		m³/h	mg/l	mg/l	mg/l	mg/l	Euro
Base DWF Set Points for Winter	1*Qin	817	1	0.5	12.7	0.6	5474
Optimum DWF Set Points for Winter	1*Qin	1857	1.3	3.3	8.0	0.8	6539
Base DWF Set Points for Summer	1*Qin	678	1	0.5	9.9	1.8	6180
Optimum DWF Set Points for Summer	1*Qin	1440	0.1	0.4	8.1	1.3	6300
Base WWF Set Points for Winter	0.5*Qin	1326	1	0.5	13.5	0.5	5763
Optimum WWF Set Points for Winter	0.5*Qin	2269	1.2	3.0	10.8	1.0	6927
Base WWF Set Points for Summer	0.5*Qin	1213	1	0.5	8.31	1.1	7462
Optimum WWF Set Points for Summer	0.5*Qin	2065	0.1	3.0	7.0	0.9	7832

DWF: dry weather flow, WWF: wet weather flow, Qrs: return sludge flow, Qir: internal recycle flow, DO4 and DO5: dissolved oxygen set points,
Ntot-N_ef: effluent total nitrogen, Ptot-P_ef: effluent total phosphorous, Cost: operational cost, Qin: average influent flow

In order to explain better the set points found in the optimization, additional operational indicators are presented in the Table 5.15. The ratio of Qir/Qin tends to indicate that the optimum set points are more dependent on the flow conditions than on the temperature conditions. From the practical point of view this may imply that a set of ratios for different influent flows may help the operators set up the internal recycle. For instance:

$$Qir = 5.3*Qin \quad \text{for DWF} \qquad \qquad \text{Eq. 5.5}$$
$$Qir = 4.1*Qin \quad \text{for WWF} \qquad \qquad \text{Eq. 5.6}$$

The oxygen transfer rate (SOTR) helps us understand the implications of the DO set points. The optimized transfer rates seem to be more dependent on the temperature conditions than on the flow. This may be explained by the transfer of gases which depend on the temperature. Since the final manipulated variable for the dissolved oxygen is in reality the air pumped, the average air flow rate (Qair) was calculated for each scenario. Qair follows two patterns: one is for winter scenarios where the air flow is slightly bigger than the base scenario, and the other is for summer where the values estimated are slightly smaller. In general, it appears that it is not the amount of oxygen transferred what helps to optimize the operation of the system but the distribution of oxygen in the reactors. From a practical point of view two rules could be derived from the optimized set points:

a. For winter season the oxygen transfer should be more evenly distributed in the two reactors with a slightly bigger rate for the first part of the reactor (R4) and

b. For summer season the oxygen transfer should be favoured in the second part of the aerated reactor (R5).

Table 5.15 Operationalization of the set points

Operational Scenario	Qir/Qin	SOTR_R4 kgO$_2$/h	SOTR_R5 kgO$_2$/h	Qair m^3 air /h
Base DWF Set Points for Winter	2.4	115	48	2119
Optimum DWF Set Points for Winter	5.5	103	90	2192
Base DWF Set Points for Summer	2.4	139	64	2487
Optimum DWF Set Points for Summer	5.1	60	84	2247
Base WWF Set Points for Winter	2.4	114	45	2052
Optimum WWF Set Points for Winter	4.1	106	90	2206
Base WWF Set Points for Summer	2.4	157	70	2851
Optimum WWF Set Points for Summer	4.1	63	157	2688

Qir: internal recycle flow, Qin: average influent flow, SOTR: oxygen transfer rate in aerated reactors R4 and R5, Qair: average air flow rate.

5.7.7 Increasing the degrees of freedom of WwTP by adding a carbon source

According to the results of the optimization for the four scenarios, the degrees of freedom available at the treatment plant are not enough to reduce the average effluent concentration of nitrogen and phosphorous simultaneously. For this reason an extra manipulated variable was explore, in this case, the addition of a carbon source at the entrance of the treatment plant. The carbon source is assumed to be a volatile fatty acid (VFA) directly added to the internal sanitary stream. In order to assess the impact of the VFA in the objectives functions, the scenarios for DWF and WWF in winter were used. The boundaries of the VFA dosing are more related to the operational cost. Thus, the boundaries of the VFA were set as VFA < 1000 mg/l for DWF and VFA < 1500 mg/l for WWF. The set points for the operational variables Qir, DO4 and DO5 were fixed with the optimum values presented above in Table 5.14.

The effects of the VFA dosing on the objective functions are presented in Figure 5.24. For both scenarios the effects are similar, with a near linear reduction in the Ntot-N and a reduction with concave down shape for the Ptot-P that tends to become asymptotic when reaching a low limit (about 0.3 mg/l).. Within the boundaries of the dosing of VFA and for these two specific scenarios, the variables have a positive effect on the removal processes. For instance, for DWF in winter, the Ntot-N was reduced from 12.7 mg/l for the base scenario to 6.2 mg/l with the optimum set points. Similar reduction factors were found for Ptot-P, which was 0.6 mg/l with the base scenario and reduced to 0.28 mg/l for the optimized set points. The reduction effects are explained by the improvement in the carbon/nitrogen ratio which facilitates the complete denitrification. In addition, the VFA is converted to PHA which may increase the phosphorous uptake ratio. The problem is that too much carbon addition can become a nuisance for the operational cost because of the chemical itself and because it has to be removed from the treatment system.

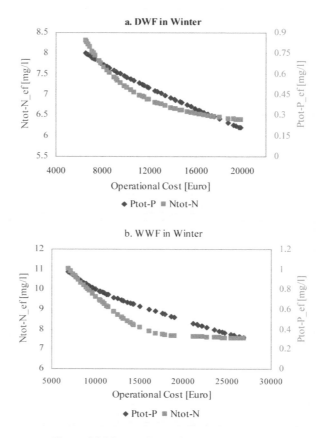

Figure 5.24 Pareto fronts for the VFA dosing

The functional design can be summarized as the selection of different set points for the operational variables (Qir, Qrs, Qair, Qex, VFA). As demonstrated above, the selection of the correct set points is a critical issue for the performance of the WwTP and this is dependent on the initial conditions of the system and the influent flows and their composition. Even with the correct set of operational strategies, the operator is left with the task of deciding when to implement them. Thus, the operator will need to define whether dry or wet weather conditions are needed and to continuously adjust the system operation for the temperature and load conditions. Therefore, this kind of "fixed set point" functional design is not the best alternative to deal with disturbances caused by precipitation events. In order to improve the disturbance rejection, it is proposed to use the underlying concepts of MoDeCo: the integration of the operation of the subsystems (sewer, treatment plant) and the sharing of information. In the case of the Gouda wastewater system, this was implemented by developing anticipatory control strategies. In what follows, an alternative is presented to improve the disturbance rejection of wastewater treatment plant and in general to improve the performance of the system.

5.8   Anticipatory Control of Gouda WwTP

The importance and usefulness of moderm meteorological data in the integrated urban water management has been demonstrated by Lobbrecht and van Andel (2005). The fast reaction of Gouda urban catchment to precipitation events ($Tc <$ hour), which is in contrast to the reaction times of biological process (from hours to weeks), supports the importance of forecasting possible disturbances at the treatment plant. To forecast the disturbance, first it is necessary to have a forecast of the precipitation. Then, with the help of the integrated model, it is possible to estimate the effects of the precipitation on the treatment plant processes. The forecast of a disturbance by itself is a tool for the operators of the treatment plant, but in combination with the optimization of the set points it can give a real improvement in the performance of the system. The implementation of the anticipatory control for Gouda UWwS and the results for selected precipitation events are presented in what follows.

5.8.1   Framework for anticipatory control of Gouda wastewater system

For the anticipatory control, the boundaries of the system are wider than those used for the functional design, because the sewer network is included as part of the system. The main disturbances of the system in this case are the precipitation events. The disturbances are propagated by the sewer model to the inflow to the wastewater treatment plant. The manipulated variables are basically the same as in the functional design, with additional degrees of freedom that come from the sewer network. Some examples of new degrees of freedom are the storage volume in the sewer network and the sewer overflows; both of the manipulated variables may influence the flows pumped to the treatment plant.

The proposed framework for the anticipatory control is shown in Figure 5.25. The diagram corresponds to a feed-forward control with an off-line optimization of the set points. To implement the framework, some additional blocks for forecasting the precipitation and disturbances were added to the code developed for the MoDeCo approach. The forecast of precipitation is used as the input to the sewer model, which produces the influent flows and the composition of the wastewater for the treatment plant. With the model of the treatment plant it is possible to see the effects at the inflow and to evaluate the need to adjust the set points. If needed, a full optimization of the set points is performed. The optimum set points in combination with the actual status of the treatment plant and the current set points generate the new control strategy for the treatment plant. The time scale for the anticipatory control is associated with the dynamics of the included processes. The time of interest for the treatment plant processes is the order of hours to days and the time scales of the sewer network processes (run-off and transport of wastewater) are also from hours to days. Since, the processes are relatively fast there is a need also to have a precipitation forecast the order of hours to days.

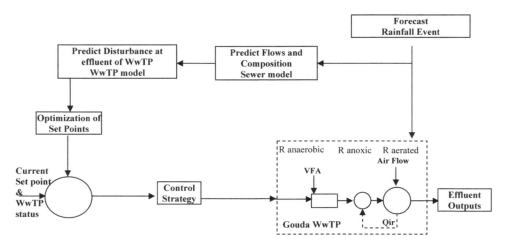

Figure 5.25 Block diagram of the anticipatory control proposed for Gouda WwTP

5.8.2 Forecast of precipitation events

The availability of the ECMWF EPS forecasts introduces new possibilities for elaborated decision support systems. The 50 ensemble members of a forecast can be used to estimate a categorical (single-valued) forecast or a probabilistic multi-value forecast (Persson and Grazzini 2007). The categorical forecast was calculated as the mean value of the 50 ensemble members, as the ECMWF manual suggests. Researchers have demonstrated that ensemble members with higher and lower precipitations have a higher probability of occurrence than do the ensemble members with average precipitation (Bokhorst and Lobbrecht 2005). Therefore, the probabilistic forecast for Gouda was estimated with the 5 and 95 percentiles of the ensemble members.

Figure 5.26 shows two examples of the 10 days ECMWF EPS precipitation forecast for De Bilt and the precipitation for Gouda accumulated every six hours. The use of the EPS forecast introduces some technical complications. Some of then are presented by Persson and Grazzini (2007) as follows:

- Use of the mean: the averaging technique works best a few days into the forecasts when the evolution of the perturbations is dominantly non-linear. During the initial phase, when the evolution of the perturbations has a strong linear element, the ensemble average is almost identical to the Control because of the "mirrored" perturbations that are added to and subtracted from the Control.
- Ensemble spread: the ensemble spread measures the differences between the members in the ensemble forecast. A small spread indicates a low forecast uncertainty; a large spread indicates a high forecast uncertainty. This also indicates how far into the forecast the ensemble mean forecast can carry information of value and helps the forecaster to determine appropriate uncertainties.
- The day–to–day inconsistency: changes in the forecast from one day to the other are necessary to enable a forecast system to take full benefit of new observations and modify previous analyses of the atmospheric state. Since the latest forecast is based on more recent data than the previous forecast, it is on average better (Persson and Grazzini 2007).

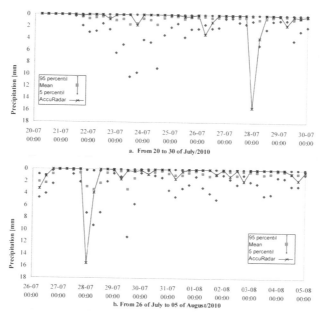

Figure 5.26 Ensemble forecast of precipitation for De Bilt and the six hours accumulated precipitation for Gouda observed by radar.

Figure 5.26a shows a good agreement between the mean and the 5 percentile precipitation forecast and the volume observed by radar. However, the forecast of the highest precipitation event at day 8 of the forecast is missed. The size of the bars between days 23 and 24 indicates a large spread in the ensemble members which means a high forecast uncertainty. Moving the window of the forecast 6 days later shows the other complication in the use of the EPS forecast (Figure 5.26b). The day to day inconsistency is clearly represented when the last 4 days of the figure "a" are compared with the first 4 of the figure "b". The forecast of the extreme precipitation observed by radar for 28 of July/2010 is different if the horizon of forecast is 8 days (in figure a) or when it is 2 days (in figure b).

The inherent complications of the use of the EPS forecast raise different questions, for instance: What is the most important information? Is it the extreme precipitation events, or are all precipitation events important? What is the forecast horizon needed? Before answering these questions it is necessary to observe how the forecasted precipitation events affect the quality of the treated wastewater.

5.8.3 Forecast of disturbance in the WwTP

To forecast the disturbance at the effluent of Gouda WwTP, the EPS forecast together with the integrated model were used. The precipitation is used as an input to the sewer model to estimate the sewer outflows. The flows are then used to compute the water quality components at the inflow to the treatment plant. With the information generated, the model of the treatment plant is run three times in order to generate the effluent concentrations for the three scenarios of precipitation (the average and 5 and 95 percentiles). Figure 5.27 shows an

143

example of the EPS forecast of precipitation and the effects on sewer outflows and the effluent of the treatment plant for Ntot-N and Ptot-P.

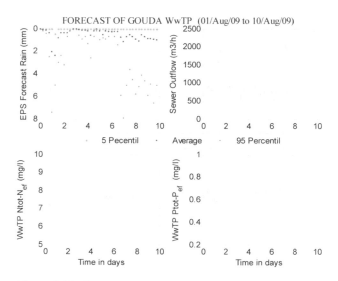

Figure 5.27. Forecast of the disturbance at the treatment plant

The EPS forecast for the example has a high spread at the beginning of the time series and at the end. The differences in the precipitation volumes are propagated by the sewer catchment outflow. However, the quality of the effluent of the treatment plant does not follow the same pattern. The system seems not to be sensitive to small precipitation events as is shown by the proximity of the Ntot-N_{ef} and Ptot-P_{ef} concentrations for 5 percentile (dotted line) and the average forecast (solid line). The capacity of the wastewater treatment plant to attenuate small peaks of precipitation may be associated with the oversized structures. The 95 percentile (dashed line) represents high precipitation events with low probability of occurrence. The peaks of precipitation raise the effluent concentrations at the beginning of the events and then lower the values after the first flush has passed and the dilution process has affected the influent wastewater. These types of events constitute a major disturbance for the treatment plant. Therefore, the most important information from the forecast of precipitation are the extreme rainfall events.

The forecast horizon may be associated with the reaction time of the process to be manipulated. As mentioned before, the reactions that can be manipulated with Qir, Qair and VFA are the order of hours to days. One of the factors that influence this decision is the lag time needed for the process to account for the changes in the manipulated variables. Judging from the results of the model this lag time for the variables mentioned above could be in the order of one to three days. Another factor that may influence the decision making based on a forecast is the experience of the institution in charge. In the case of Rijnland Water Board, they have experience with decision support system that use one day of forecast. Therefore the horizon of the forecast could be around three days.

The use of the forecast should be supported by an analysis of the accuracy of the forecast. The analysis should include the accuracy of the precipitation volumes and the peaks of precipitation. An analysis of the spatial resolution of the forecast should also be considered.

Since the propagation of the disturbance cause by precipitation is affected by the status of the sewer network and the treatment plant, the accuracy of forecasted flows and water quality composition should also be analyzed. Utility aspects like the success in the use of the forecast in the reduction of operational cost or water quality impacts should also be carried out. The accuracy and utility analyses are out of the scope of this research. However, a detail analysis of the accuracy of the forecast with respect to measurements of precipitation can be found in Andel (2009). According to Andel, *et al.* (2008) results, the ECMWF EPS precipitation forecast can be used to forecast critical events for the area of Rijnland Water Board. Even more, his research suggest that to identify critical events, a low probability thresholds (<0.05) should be used; in other words the 95 percentile probabilistic forecasts. Consequently, it is assumed that for the anticipatory control of the WwTP of Gouda, the 95 percentile probabilistic forecast can be used to identify critical events and to optimize the performance of the system.

5.8.4 Rejection of disturbances by optimizing the set points

The rejection of disturbance required the integration of all the pieces of code developed and the knowledge acquired from the data of the system. The forecast of precipitation and the effect in the effluent composition of Ntot-N and Ptot-P presented in the Figure 5.27 are used to illustrate the rejection of disturbances. Following the previous discussion, the 95 percentile forecast is used to illustrate the risk of critical precipitation events. The 10 days period corresponds to a summer period with wet weather conditions. The time series of precipitation is used in the sewer model to produce the influent flows and their compositions to be modelled in the treatment plant. With the estimated wastewater characteristics and using the functions of Equation 5.1, the optimization algorithm found the Pareto solutions shown in Figure 5.28a. The values of the functions generated with the base set-points and the selected set points for the functional design are also plotted in Figure 5.28a. With the heuristic rule proposed previously, one of the optimum solutions is selected for further analysis. The dots inside the rectangular shape in Figure 5.28a correspond to the selected solution.

The effect of the three operational set points in the time series of effluent nitrogen and phosphorous is presented in Figure 5.28b. The optimization algorithm found a solution that is better that the base scenario and better than the optimum set points for the functional design. The effluent concentration is better for nitrogen and slightly worse for phosphorous, being consistent with the previous analysis. The set points for each scenario of the variables are compared in Table 5.16. Giving priority to the removal of nitrogen over the removal of phosphorous and the operational cost, the optimum set points are those found for the specific disturbance that is affecting the treatment plant.

Table 5.16 Comparison of the objective functions for the basic and optimum set points

	Set Points			Optimization Objectives		
	Qir	DO4	DO5	Ntot-N	Ptot-P	Oper Cost
	[m3/h]	[mg/l]	[mg/l]	[mg/l]	[mg/l]	[Euro]
Base Scenario	1161	1	0.5	7.5	0.77	7869
Functional Design	2065	0.1	3	6.0	0.73	8172
Optimum Disturbance	3113	0.2	1.6	5.1	1.03	9001

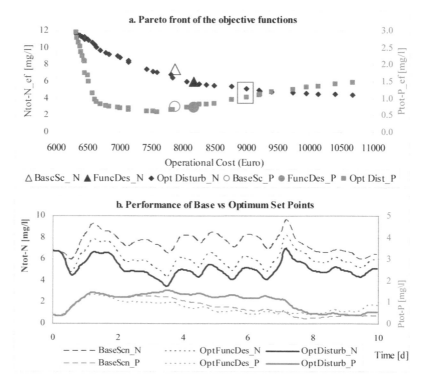

Figure 5.28 Solutions of the optimization of the set points. a. Pareto sets b. Performance of selected solution compared with the basic set points.

As a result of this experiment it may be concluded that for disturbance rejections it is important to optimize the set points specifically for the event. The fixed set-points of the functional design are not enough to handle critical rainfall events. Therefore, the implementation of an anticipatory control system for Gouda may contribute to improve nutrient removal efficiencies and disturbance rejections.

5.9   Discussion

In general, the results of the experiments show that is possible to improve the performance of Gouda UWwS by optimizing the functional design. The functional design was summarized as the selection of set points for the operational variables: internal recycle (Qir), dissolved oxygen in two sections of the aerated reactor (DO4, DO5) and dosing with the carbon source (VFA). With these manipulated variables, it was possible to reduce the effluent concentration of total nitrogen while keeping the concentration of total phosphorous within set boundaries. For instance, for DWF in winter, the concentrations of Ntot-N and the Ptot-P were reduced by 51% and 53% respectively when the performance of the system with optimized set points is compared with the base scenario defined above. The performance of the system with respect to operational cost decreased, but that was expected because the objectives are contradictory (i.e. decreasing the effluent concentrations implies an increase in cost). However, the estimated costs do not include the possible cost savings associated to the reduction of pollution impacts in the river. Therefore, the cost function values must be regarded as order of magnitude values rather than exact values. In general, further resources should be put into reducing the uncertainty in the models used to produce the function values, that is, to reduce the uncertainty in the models of the sewerage network and the treatment plant and to improve the analysis of costs used to compare alternatives.

The evaluation of multiple combinations of Qir and DO set points shows that a selected combination may reduce the effluent concentration of nitrogen but it also leads to an increase of phosphorous. This contradictory behaviour is associated with what Olsson and Newell (1999) term internal disturbances. Therefore, the design of the operation of the treatment plant must consider the system as a whole rather than the optimization of the operation of the sub-processes (e.g. nitrogen removal) separately. In other words, the functional design should find the set points that optimize the process of interest with out negatively affecting other processes in the system. It should be noticed that this analysis was concentrated only on variables that manipulate the nutrient processes within hours to days. However, further research should include manipulated variables that influence slow processes like the growth of microorganisms. A full functional design should also include state variables within the processes and longer operational horizons, for instance including a period that allows the evaluation of changes in the sludge retention time or sludge production (i.e. two or three months).

The paradigm of functional design using fixed rules or set points has to be changed. The results of the experiments indicate that a functional design based on "fixed set point" is not the best alternative to deal with disturbances cause by precipitation events. The set points are dependent on two highly dynamic factors: the current status of the system and the influent flow and its composition. Therefore, it may be possible to find optimum set points for each specific operational situation. In practice, the treatment plants rely on the expertise and experience of the operators to set up the control strategies to deal with highly variable influent flows and compositions. To avoid fixed rules and facilitate the decision process of operators, the anticipatory control strategy seems promising. However exploratory, the results show that it is feasible to predict the disturbances created by precipitation events and by using the optimization algorithms, it is possible to improve the disturbance rejection capacity of the system. Further research in this area should be focused on the verification of the prediction of the disturbances, considering the propagation of the effect of the precipitation through the treatment plant. In addition, utility aspects like the cost saving made by the use of anticipatory control strategies should be explored.

Within the main benefits of the use of MoDeCo approach to design the functional strategy of Gouda it is possible to highlight:

- The generation of new knowledge about the behaviour of the system. For instance, the ratio of Qir/Qin tends to indicate that the optimum set points are more dependent on the flow conditions than on the temperature conditions. From the practical point of view this implies that a set of ratios for different influent flows may help the operators to set up the internal recycle. Another key finding was that the oxygen transfer rates seem to be more dependent on the temperature conditions than on the flow conditions. In general, it seems to be that it is not only the amount of oxygen transferred what helps to optimize the operation of the system but also the distribution of the oxygen in the aerated reactor. The knowledge generated may contribute to the improvement of the operation of the system and the reduction of pollution impacts on the receiving system.
- Another benefit arising from the approach proposed is the possibility of an integrated analysis of the system, for instance, the use of the plant wide model to understand the

147

effect of operational variables on all processes involved in the treatment, or the integration of the sewer system into the anticipatory control of the Gouda WwTP to help identify disturbances created by precipitation events.

- Contrary to the actual design of Gouda based on fixed rules selected for steady state conditions, the approach allows us to consider the dynamics of the system. For instance, the derived set points are based on 10 days of diurnal variations of flows and composition load.

- The use of optimization algorithms increases the chance of finding optimum solutions that consider multiple variables and multiple objectives. The search space of the variables is so wide that it will be almost impossible to explore it in an exhaustive search and impossible to evaluate it in a scale model in a laboratory or in the real treatment plant.

The main limitations found for the application of MoDeCo approach to the functional design are:

- Advanced model based design may require more information of the system than the traditional design approach. The information available for the construction of models is normally limited. The limitations on the information may affect the accuracy of the models used. However, the case of Gouda proves that with the current operational monitoring system it was possible to develop behavioural models of the treatment plant.

- The computing time may be a limiting factor if the analysis horizon is extended. For instance, the computational time in the analysis of slow reaction processes (microorganism growth) may be prohibitive. For the anticipatory control, on-line optimization will not be an option because the basic running time for ten days of analysis is about four hours. To improve the MoDeCo approach, further research should be oriented to reducing the computational time needed for the optimization processes.

- The Pareto solution set is not composed of only a single point so that the selection of the desired point requires information regarding the ranking of alternatives. However, defining the function weight (preferences) a unique solution may be selected. The fact that multi-objective optimization generates a set of solutions limits the use of the tool for on-line automated optimizing set points. However, for off-line optimization it is an excellent tool because it gives the opportunity for the operator to be part of the decision processes.

- The holistic view and the integration of modelling tools require certain skills and knowledge that may limit the application of the methodology. This kind of approach requires interdisciplinary work and the will of the institutions in charge of the functional design to use advanced Hydroinformatics tools. The transfer of knowledge in this field may help to reduce possible resistance of the design engineers to the use of mathematical tools for optimization.

5.10 Conclusion

The approach proposed contributes to the improvement of the removal efficiencies of nutrients at the treatment plant of Gouda by the optimum selection of the operational set points. The analysis based on modelling tools brings the possibility of including the dynamic behaviour of the influent flow and composition. The significant number of scenarios analyzed allows the designer to understand better the system. The learning processes go beyond the set points and help us to understand the interactions between processes and the influence of the control variables. The integrated model also contributes to understanding the synergy between components (sewer, treatment plant). Even though, there is a clear benefit the outcomes are to be seen as qualitative and not quantitative. For the application of the operational set points to a real system, an important effort has to be done in validating the models and the results obtained with them.

The introduction of the ECMWF EPS forecast of precipitation may be a great opportunity to create decision support systems for the operation of wastewater treatment plants. This idea seems to be promising and will help to reduce the need to fix operational rules in a reactive way. An anticipatory control will help prepare the system to deal with more frequent and intense peaks of flow and pollution cause by changes in the climate or the reduction of CSO operation in the urban catchment. Further research into the validation of the forecast and the implications of the implementation of the decision support system will greatly contribute to the improvement of disturbance rejection in treatment plants and thus to pollution control in urban wastewater systems.

6 Use of Cloud Computing and Surrogate Modelling in Optimization Processes[2]

6.1 Introduction

One of the main limitations found during the application of model based design and control (MoDeCo) approach was the long computing time required during the optimization processes. The application of Multi-Objective Evolutionary Algorithms (MOEA) in chapters 4 and 5 shows the high potential of this type of algorithms. However, many authors have recognized that the long computing time required to identify solutions makes the approach less attractive or even infeasible for practical applications (Coello Coello, *et al.* 2002, Schütze, *et al.* 2002, Vanrolleghem, *et al.* 2005, Fu, *et al.* 2008). Therefore the motivation of this chapter is the need to reduce the computing time in MOEAs, in order to make the MoDeCo approach applicable to complex urban wastewater problems.

The problem with the long computing time has different reasons and it has been addressed from different angles (Figure 6.1). First of all, in the optimization of a UWwS we do not have an analytical expression of the objective function but can only calculate it by running a complex model which requires considerable computation time. This means that it is not possible to use efficient gradient-based algorithms and therefore one should apply so-called direct optimization. A large class of such algorithms is randomized search, and MOEA belongs to this class of algorithms. Secondly, optimization is performed in the space of many decision variables (called the search space, or design space) so inevitably, even with efficient optimization algorithms the number of function evaluations (and hence model runs) makes the computing time prohibitively long. An approach to solve the problem is the parallelization of the algorithms in combination with distributed computing power (Martins, *et al.* 2001).

In the optimization of real-life water systems we have to deal with complex process models that are used to calculate the objectives function(s). For instance, flow routing in a sewer network requires the solution of the Saint Venant equations, which implies the use of complex algorithms, and this has to be taken into account in the design of the optimization process. Thus a second approach to reduce computing time is the use of surrogate models (known as approximation models, metamodels or response surface models) that mimic the mechanistic model but are computationally less demanding.

To the author's knowledge there are not as yet any known studies comparing the above two approaches and deducing their efficacy to reduce the computing time for a practical optimization problem. Therefore, one of the main contributions of this chapter is the comparison of the use of surrogate models and parallel computing in multi-objective optimization. Furthermore, a novel approach that uses virtual clusters in the Cloud as a parallel computing infrastructure is compared with the traditional cluster form with networked

---

[2] This chapter is partially based on the results of the MSc research of Xu Zheng. The author mentor the student under the supervision of Prof Dimitri Solomatine, from the core of Hydroinformatics at UNESCO-IHE Institute for Water Education, and the support of Dr. Francesca Pianosi from the Dipartimento di Elettronica e Informazione, Politecnico di Milano, Milan, Italy.

workstations. In addition, a surrogate modelling approach based on the ideas presented by Liu, et al. (2008) is developed. The resulting algorithm is named Multi-objective Optimization by PRogressive Improvement of Surrogate Model (MOPRISM). The design of an urban drainage system in Colombia is used to test the approaches. The design of an urban wastewater system is a large scale multi-objective optimization problem using a complex hydrodynamic model, so it was seen as a good case study. Results are discussed and compared with those obtained by a standard NSGA-II optimization algorithm. The limitations of the methods and ideas on its further improvement are presented as well.

Figure 6.1 shows the methods used to reduce computing time; the alternatives tested in this chapter are in bold letters.

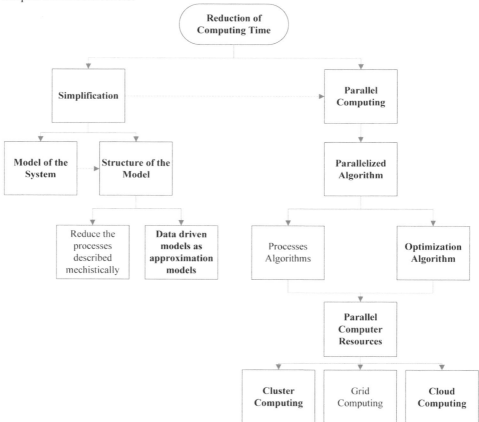

Figure 6.1 Alternatives to reduce the computing demand of optimization processes

6.2   Parallel Computing Optimization

Following the development of multi-core processors and the ability to connect computers together in clusters or grids, parallel computing techniques have emerged as an alternative to speed up the computation of optimization problems (Martins, *et al.* 2001). However, parallelism leads to the need for new algorithms, specifically designed to run simultaneously on different processors, together with the parallel computer resources suitable to run the parallelized algorithms. Parallel computing has traditionally been done on expensive mainframe computers that require skilled support personnel. More recently, clusters of standards computers connected by Ethernet have become widely used (Beowulf clusters), but a dedicated cluster requires significant time and effort to construct and maintain. A variation of dedicated clusters is the use of Networks of Workstations (NOW) that operate part time as a cluster. A NOW cluster depends on the availability of idle workstations, and the speedup of the parallelization can be reduced as a consequence of the migration of jobs when a machine is no longer available.

An alternative to satisfy the demand of computer resources, is *Cloud computing*. Cloud computing and its inherent ability to exploit parallelism at many levels has become a fundamental new enabling technology to facilitate the access to computational capabilities for parallelism users (Gannon and Reed 2009). The facility to scale up the computer infrastructure on demand can bring important advantages to make parallel computing both easier and more desirable.

6.2.1 Parallel computing algorithms

In the optimization process for a UWwSs there are two algorithms than can be parallelized: the algorithm that simulates the processes in the urban catchment, and the optimization algorithm. The parallelization of algorithms for the hydraulic computations for sewers and rivers require considerable changes to the existing codes and are quite limited. A recent attempt to parallelize a simplified conceptual model for an integrated UWwS is presented by (Burger, *et al.* 2009). The maximum speedup achieved was 4.2 times in eight threads, which is in accord with the complexity of parallelizing a fine grained code with many interdependencies.

In contrast to UWwS models, many algorithms used for model-based optimization can be easily parallelized. MOEAs are frequently used in the optimization of a UWwS and they are very suitable for parallelization, because the objective function evaluation can be performed independently on different processors. Three approaches are found for parallelization of EAs: Master-slave, Island model and Diffusion model (Branke, *et al.*). The Master-slave approach is frequently used because of its simplicity; a single processor (master) maintains control over selection, cross over and mutation, and uses the other processors (slaves) for the evaluation of individuals (Cantu-Paz 1999).

Barreto, *et al.* (2008) parallelized the Non Sorted Genetic Algorithm (NSGAX) using the master-slave approach to find optimum solutions for urban drainage rehabilitation. For an experiment with 4 computers the speedup reported by the authors was 2.6, and, as shown by the latest experiments, for more computationally demanding models the speedup was higher. As expected for this kind of parallelization, the efficiency is reduced as the number of processors is increased. According to Cantu-Paz (1999), master-slave GAs have frequent inter-processor communications and it is likely that the parallel GA will be efficient only for problems where the fitness function evaluation is computationally demanding. The reason for this is that as more slaves are used the time used for communication between processors increases. Even though there are more efficient parallelization algorithms like the island model or the cone separation (Branke, *et al.* 2004), the master-slave approach seems to be a good alternative to test parallelism on the cloud.

6.2.2 Parallel computer architecture and cloud computing

Master-slave approach requires computing machines that are distributed and may form clusters; this architecture is briefly presented below.

Cluster

A cluster is a local group of networked computers with installed software that allows them to work simultaneously in parallel. Clusters for parallel computing require a high-speed, low-latency network in order to achieve high performance. Latency refers to the time it takes for one processor to communicate with another. Key features are the bus speeds that connect the CPU to memory, power consumption per CPU, and the networking technology that connects the CPUs to one another (Creel and Goffe 2008). If NOWs are used then there is no need for the physical creation of the cluster but still installation of the parallel communication software is needed. Reliability also depends on the network connection and the availability of idle workstations.

Cloud

Cloud computing as defined by U.S. National Institute of Standards and Technology (NIST) is an environment for enabling on-demand network access to a shared pool of configurable computing resources that can be rapidly provisioned and released with minimal management effort. This type of Cloud service called Infrastructure as a Service (IaaS) offers the possibility to create a full computer infrastructure (i.e. virtual computers, servers, networks, etc). Through virtualization technology and parallelism users can create a "Virtual Cluster" dedicated and customized for the problem in hand. The scalability of the infrastructure in the cloud facilitates the increase or decrease in the computational capacity on-demand (Mell and Grance 2011). Since this is a pay-as-you-go service, no maintenance cost or information technology personnel are required. Images with customized software can be created and used to facilitate the instantiation of virtual computers. However temporary run-outs of capacity by providers may diminish the reliability, latency may reduce computation efficiencies, and internet connection bandwidth may limit the front end users.

6.3 Surrogate Model Based Design Optimization

Surrogate models have been used in multiple applications to approximate computationally demanding process based models (Solomatine and Torres 1996, Maskey, *et al.* 2000). The most common approach uses surrogate models in sequential mode. In a sequential approach (Figure 6.2a), first the surrogate model is constructed based on a selected sample of the search space, then the optimization process evolves using the surrogate model until an approximate optimal Pareto frontier is obtained that fills the convergence conditions, and at last the Pareto solutions are verified with the mechanistic model (Wang and Shan 2007). The problem with this approach is that the simplification of the mechanistic model leads to a loss of accuracy in the representation of the system being modelled. Therefore, in the optimization process with surrogate models there is no guarantee of obtaining the same solutions as when using complex mechanistic models (Liu, *et al.* 2008).

In order to reduce the loss of accuracy of the subrogate models, a different approach has evolved. The main strategy consists in updating the subrogate model in a loop within the optimization process. Nain and Deb (2002) proposed an approach combining a genetic algorithm with artificial neural networks (ANN) as an approximation technique. The approach starts with the computationally demanding model for a number of generations, then the ANN model is trained using a set of samples from solutions previously found. Then a fixed number of generations are evaluated using the ANN model as surrogate model, and this process cycle is repeated until the convergence criteria is fulfilled (Figure 6.2b). As pointed out by Gaspar-Cunhaa and Vieira (2003), one of the critical issues for this approach is to define the number of generations to evaluate the computationally demanding model and the surrogate model. They proposed an improvement in the method by introducing a measurement of the error produced by the ANN approximation in each generation, thus eliminating the need to define and fix the number of generations that should be left to evolve. In a second hybrid approach they use the ANN first with a local search algorithm to find some tentative solutions.

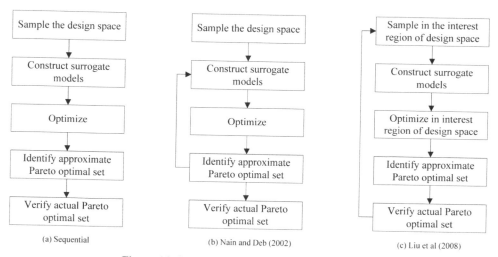

Figure 6.2 Surrogate based optimization schemes.

Another hybrid approach called ParEGO was proposed by Knowles (2005). ParEGO uses one of the kriging approaches, the Design and Analysis of Computer Experiments (DACE), as an approximation method. In one of the few reported applications of the hybrid approach to urban wastewater systems, Fu, et al (2008) tested ParEGO versus NSGAII to optimize real time control strategies. For that application, ParEGO found the same Pareto solutions with 260 evaluations of the objective function as did NSGAII after 10000 evaluations. The limitation with this approach arises from the fact that it aggregates the objectives in a single objective, thus it requires the definition of the weights for each objective. In order to explore the whole region of the Pareto front, a varying weight vector has to be used which may increase the number of evaluations required. Liu, et al. (2008) pointed out that all the methods presented above try to minimize the loss of accuracy of the surrogate model, but the evolution of the optimization still depends significantly on the accuracy of the surrogate model. In order to reduce the dependency they include a variation in the approach in which not only the surrogate model is updated but also the interest region of the design space. With this variation they found that the algorithm searches for more solutions near to the Pareto set even with less accurate models.

6.3.1 Surrogate models

The use of surrogate models for the optimization process implies the selection of a particular type of model. One alternative is to simplify the mechanistic models. For instance, Schütze, *et al.* (2002) use a series of linear reservoirs to model flow within urban sub-catchments and in that way simplify the algorithm for routing the water in the system. In a similar application, Meirlaen, *et al.* (2001) use tanks in series to represent a river as part of an integrated model and thus reduce the computing time in this way. However, the reduction of time is limited by the degree of simplification of the mechanistic model, and that depends on the problem in hand.

Perhaps the most used type of surrogate models are data-driven models due to their low computing time requirements. For instance, Lobbrecht and Solomatine (2002) demonstrated the advantages in computing reduction using artificial neural networks (ANN) to optimize control strategies in drainage systems. Since the objective of this chapter is the reduction of computing time, data driven models appear to be a good alternative as surrogate models. The accuracy of this type of model depends significantly on the determination of data sets from which the algorithm will learn and the selection of the learning algorithm. Supervised machine learning takes a known set of input data and known responses to the data, and seeks to build a predictor model that generates reasonable predictions for the response to new data. Normally the learning theory is based on the randomly selected training and testing data set, so that the algorithm is considered as passive and has no control over the information that it receives (Freundy and Seung 1997). In contrast, active learning algorithms are allowed to ask "questions" in order to accelerate the learning process (Liere and Tadepalli 1997). The selection of the learning algorithm from a group of methods on a given data set should follow the so-called Occam's razor principle (Mitchell 1997) which states that if two models return the same level of accuracy then the simpler one is always preferred.

6.3.2 Function of subrogate models

In the optimization process surrogate models can play three different roles: to approximate the process-based models, the objective functions or the Pareto frontiers (Figure 6.3). In the first option (link 1) data driven techniques are used to build a model of models. This use of the surrogate model has been widely employed since it can obtain a good representation of the modelled system (Solomatine and Ostfeld 2008). However in urban wastewater systems model outputs are variable in space and time which may make them very complex to approximate. In contrast, the third option provides a direct bridge from the model input to the Pareto solutions (Link 3); the jump reduces the overall computational time, but is highly dependent on the accuracy of the approximation. Perhaps in a more efficient role the surrogate models can be used to explore the relationships between several explanatory variables and one or more response variables (Link 2). This method is also known as the Response Surface Methodology (RSM). Few applications have been found for this method applied to water systems; see for instance Castelletti, *et al.*(2010) who use the response surface strategy to optimize the design of an aeration system for a reservoir.

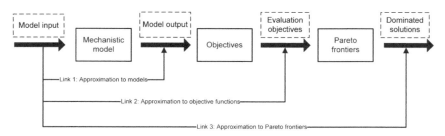

Figure 6.3 The contrast between often used surrogate model and response surface functions

6.4   Formulation of the Problem and Objectives

Problem statement

The long computing time required by multi-objective optimization algorithms to identify the Pareto solutions makes the approach less attractive or even infeasible for practical design applications. Therefore, there is a need to find alternatives to reduce the computing time of the optimization processes in order to make the MoDeCo approach applicable to complex urban wastewater problems.

Objectives

The main objective is to contribute to the existing knowledge of alternatives to reduce computing time in optimization processes. The specific objectives are:

- Evaluate the reduction in computing time using parallel algorithms and parallel computer infrastructure for the optimum design of a UWwS.
- Compare two types of parallel computer infrastructure: NOW cluster and cloud computing cluster.
- Evaluate the reduction in computing time using surrogate modelling for the optimum design of a UWwS.
- Compare different types of data driven models used as surrogate models.

Aim of the chapter

The aim of this chapter is to give UWwS designers who want to use multi-objective evolutionary algorithms an overview of possible benefits and drawbacks of parallel computing and surrogate modelling in dealing with computational demanding problems.

6.5   Data and Methods

The problem that was used as a case study corresponds to the optimization of a sewer network for the expansion zone of Cali – Colombia. Details of the case and the models used to simulate the UWwS are presented in Chapter 4. The sequential optimization based on the mechanistic model to estimate the functions values is called here the standard optimization process (Figure 6.4 a). The standard optimization process was used to define the base line for the computational time. Two strategies were implemented to reduce the computing time of the optimization process. The first strategy implemented was parallel computing (Figure 6.4 b). For parallel computing, the computational demanding process based models were used in combination with the parallelized optimization algorithm. Two different parallel computer infrastructures were tested; cluster and cloud computing.

The second strategy to reduce computing time was the simplification of the model structure (Figure 6.4 c). Four different data-driven models were tested as surrogate models: Linear Regression (LinearR), Artificial Neural Network (ANN), Regression Tree Models (RTree) and k-Nearest Neighbour (k-NN). The surrogate model was used with a sequential optimization algorithm in a single processor. Computing times were compared using various performance indicators. The details of the methods used are presented in what follows.

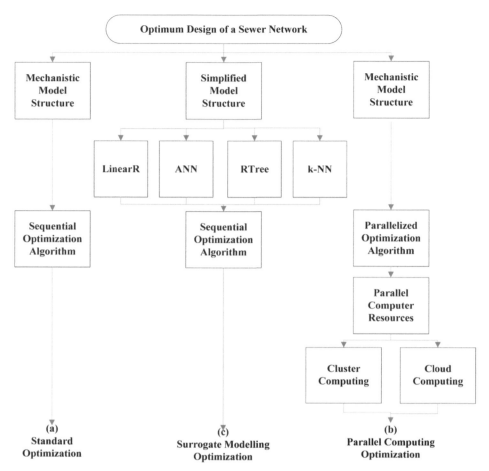

Figure 6.4 Methods to Reduce Computing Time in Optimization Processes

6.5.1   Methodology for multi-objective optimization using parallel computing

The framework for optimization processes using parallel computing is presented in Figure 6.5. The Master-Slave approach was used for the parallelization. The method requires three components: the parallelized optimization algorithm, the parallel computers and communication software and the model of the system to be optimized (Figure 6.5). Each component is explained below.

Parallel multi-objective optimization algorithm
Until now we have being using the multi-objective optimization algorithm NSGA-II (Deb et al., 2002), implemented in the Genetic Algorithm Toolbox of MATLAB, to generate the Pareto front solutions. MATLAB also provides a parallel computing function for the NSGA-II; therefore there is no need for changes in the code. The Parallel Computing Toolbox can automatically distribute computations to multiple processors, so the parallelization follows the master-slave approach.

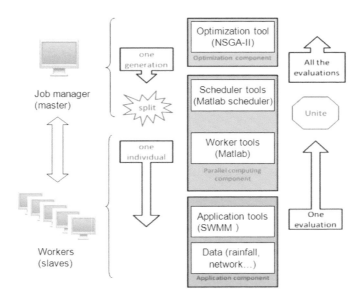

Figure 6.5 Framework of optimization of urban wastewater system using parallel computing

Source: (Xu, *et al.* 2010)

Parallel computer and communication library

As parallel computer infrastructure, two types of clusters where built: a "local cluster" of workstations from the local network of UNESCO-IHE Institute for Water Education, and a "virtual cluster" using as a service provider Amazon's Elastic Compute Cloud (EC2). Amazon machine images (AMIs) of the system's configuration including applications, data and libraries were created and stored in Amazon S3. The AMIs were used to duplicate instances in the virtual cluster. To establish parallel communication between the PCs in the clusters, the Parallel Computing Toolbox functions, MATLAB Distributed Computing Engine (MDCE) and the scheduler were used.

Performance indicators for parallel computing

The speedup, efficiency and improvement were used as indicators to compare the results of the parallelization using different clusters. Speedup is the fraction of execution time for instructions in sequence to the execution time for instructions in parallel. Efficiency describes the performance of computation system and is calculated as the speedup divided by the number of instances.

6.5.2   Methodology for multi-objective optimization using surrogate modelling

The speedup, efficiency and improvement were used as indicators to compare the results of the parallelization using different clusters. Speedup is the fraction of execution time for instructions in sequence to the execution time for instructions in parallel. Efficiency describes the performance of computation system and is calculated as the speedup divided by the number of instances.

In general the framework proposed in Figure 6.6 follow seven steps:
1. Randomly generate population X_0.
2. Pass X_0 to data-driven modelling module as $X_i^{samples}$
3. Evaluate $X_i^{samples}$ in the mechanistic model to find out $f_i^{mechanistic}$.
4. Use $X_i^{samples}$ and $f_i^{mechanistic}$ to train the surrogate model F_i, where F_i contains all objective functions.
5. Use surrogate model and initial population $X_i^{initial}$ for the optimization algorithm for k generations, and pick up the last population of last generation X_k
6. Pass $X_i^{Non-dominated}$ to data-driven modelling module.
7. Return to step 3 until finishing the number of iterations specified

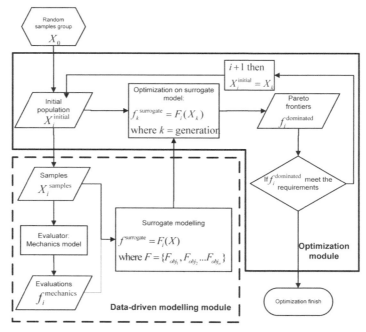

Figure 6.6: Framework of surrogate methodology

Source: (Xu, *et al.* 2010)

Data driven modelling module for surrogate optimization

The modelling module was developed in MATLAB. This module creates and updates the data driven model used as surrogate model in the optimization process. The creation of the surrogate models for optimization follows the supervisory learning process:

- Sampling the search space: the method proposed for sampling starts as a random sampling within the boundaries of the variables, and after the first iteration the sampling is updated with the variables that correspond to the non-dominated solutions of the Pareto. Therefore the sampling follows the search pattern of the NSGAII algorithm (Deb, *et al.* 2002). The size of the sample is limited by the computing time required to evaluate the mechanistic model. If the sample is too small the data will not be enough to train a surrogate model, but if the sample is too big, the computing time will reduce the efficiency of the method proposed. Therefore a sample size of 100 was used for the experiments.

- Prepare data: a vector of sampled variables (X_i) is used as input in the mechanistic model to create the response function values $(f_i^{mechanistic})$. For the problem addressed here the response function values correspond to post-processed outputs of the mechanistic model.
- Selection of learning algorithm: Because the accuracy of the surrogate model depends on the learning algorithm selected, four algorithms were tested: Linear Regression, Artificial Neural Network, Regression Tree and k-Nearest Neighbour. The algorithms used for the experiments correspond to the implementations available in the Statistics tool box and Neural Network tools box of MATLAB 2009.
    - Linear Regression models (LinearR) represent the relationship between a continuous response f_i and a continuous or categorical predictor X_i as a linear combination of (not necessarily linear) functions of the predictor, plus a random error ε. Since the sampled vector of variables (X_i) is multidimensional, so are the functions f_i that form the terms of the model.
    - k-Nearest Neighbour (k-NN) is a classification method in which the target function values (i.e. flood volume) for a new vector of sampled variables (X_i) is estimated from the known values of the k nearest training examples. The algorithm includes the following characteristics: i) The euclidean distance was used to locate the nearest neighbours; ii) the decision rule to derive a classification was a weighted sum of the k-nearest neighbours' function values. The weights were estimated based on Euclidean distances. iii) The number of neighbours used to classify the new sample was k = 5.
    - Regression Tree (RTree): RTree is a binary tree where each branching node is split based on the values of the input vector (Xi). The algorithm computes the full tree and the optimal sequence of pruned sub-trees. The minimum number of observations for a node to split was selected as sp = 10.
    - Artificial Neural Network (ANN) is a robust learning method to approximate non linear response functions. The ANN used can be described as a Multi-layer network trained with the backpropagation algorithm called trainlm in MATLAB. The architecture of the network is fixed and includes three neurons, a hidden layer with a hyperbolic tangent sigmoid transfer function (tansig) and an outer layer that uses a linear transfer function (purelin). The number of iterations for the learning process was fixed at 50.

- Fit the model: input vector (Xi) and response function values $(f_i^{mechanistic})$ constitute the training data. A supervised learning algorithm analyzes the training data and produces an inferred function, which is called here the surrogate model (F_i).

- Validate the fitted model: the root mean square error (RMES) was used as the error estimator to compare the fitted models. The error of the fitted model was calculated using cross-validation (Wang and Shan 2007).

- Pass the fitted model to the optimization module: in this step the surrogate model as produced by the fit model function is passed to the optimization module. In the optimization module the surrogate model is used to predict the response function values $(f_k^{surrogate})$ of the new input variables (X_k).

- Update the surrogate model: the model is updated using a new sample of data (Xi = Xk) that corresponds to non-dominated solutions found in the optimization processes.

Optimization module
The optimization module was developed in MATLAB. This module controls the iterative process of the optimization, runs the optimization algorithm using the surrogate model and creates the new sample to update the surrogate model. The multi-objective optimization algorithm corresponds to the NSGA-II (Deb et al., 2002), implemented in the Genetic Algorithm Toolbox of MATLAB. Additional code was written in MATLAB to adapt the algorithm to the needs of the module and to create the new sample of variables ($X_i^{samples}$) after each of the iterations. Details of NSGA modifications are presented in Chapter 4. The parameters used in the optimization processes are presented in Table 6.1.

Table 6.1 : Parameters Used in the Optimization Process

Parameters of optimization	Model used in optmization				
	Mechanistic	Surrogate			
	UWwS model	LM	ANN	RT	k-NN
Population size	100	100			
Generations	20	25			
Stopping rule	20 generations	10 iterations			

Performance indicators of surrogate optimization process
Even though the objective of these experiments is to evaluate the reduction in the computing time, we also need to assess the quality of the Pareto solutions obtained with the surrogate models. Previous researchers have used the number of exact evaluations of the objective functions as the significant running parameter, neglecting the computing time required to train and test the surrogate model (Gaspar-Cunhaa and Vieira 2003). Considering that in computational demanding problems the previous assumption is valid, we use the number of function evaluations in the mechanistic model (Fe_m) as the criteria to asses computing time. The reduction of computing time is evaluated by comparing the number E_m required to reach a "similar quality" of the Pareto solution set evolved with 20 generations with the mechanistic model.

The quality of the Pareto solution set depends on a number of factors which include the closeness of the points obtained to the True Pareto Frontier (TPF), the number of points obtained, and how well the points are distributed on the Pareto frontier (Khokhar, *et al.* 2010). The performance indicators should enable us to monitor the quality of a Pareto solution set as obtained by a multi-objective optimization method, and compare the quality of Pareto solution sets found by different multi-objective optimization methods. A important number of quality indicators have been developed to evaluate Pareto solution sets; see Azarm and Wu (2001) . From the possible metrics we select two to be used in the experiments: Hypervolume and Pareto Spread.

Hypervolume: is also called hyperarea metric or S metric and is one of the most used indicators to compare the outcome of MOEAs (Bader and Zitzler 2008). Hypervolume (HV) can be defined as the volume in the objective space covered by members of the non-dominated set of solutions. For a two objective function case, the HV can be described as follows: for each solution f_i, a hypercube v_i is constructed such that the solution f_i and a reference point R are its diagonal corners of the hypercube (Figure 6.7). The reference point can be defined as a vector of the worst objective function values. Then, the HV is estimated as the union of all hypercubes generated (Eq. 6.1). This measure does not need previous knowledge of the TPF because it focuses on measuring the dominated space. Therefore, when comparing two Pareto solution sets, the best one is the one that has larger values of HV. For the calculation of the HV, the set of Pareto solutions were normalized (between 0 and 1) and the vector [1, 1] was used as the reference point R There are different method to calculate HV as described by Anne et al (2009); here we approximate the values using the Monte Carlo method. The used algorithm corresponds to the implementation of the Monte Carlo method by Cao and can be downloaded from MATLAB Central (Cao 2008).

$$HV = \bigcup_{i=1}^{k} V_i$$

Eq. 6.1

Where: HV: union of all hypercubes
V_i : hypercubes

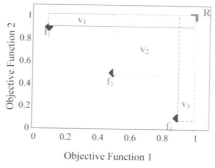

Figure 6.7 Illustration of hypervolume in a Pareto front

Spread: This indicator is used to quantify how well the points are distributed on the Pareto frontier and is widely used to compare the performance of MOEAs (Okabe, *et al.* 2004). The estimation of the spread (S) is based on the method described by Deb et al., (2002). S is calculated as the average crowding distance of the solutions in the non-dominated solution set (Figure 6.8). When comparing two Pareto Solution sets, the one having a smaller indicator is better because this indicates a better spread of the solutions in the front. The algorithm used to estimate the S is based on the implementation of Crowding Distance in MATLAB (2009).

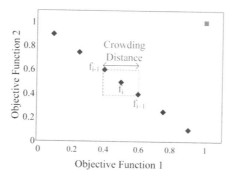

Figure 6.8 Illustration of crowding distance in a Pareto front

6.5.3 Case study – optimum design of a sewer network for Cali - Colombia

As a case study for the research, the design problem of an urban drainage system for Sector 1A of the expansion zone of Cali (Colombia) was implemented. The UWwS covers an area of 70 ha and will provide service for 22000 inhabitants. It is composed of a combined sewer network with an online storage tank, a combined sewer overflow (CSO) and the WwTP. Effluents of the system are discharged by gravity to the Lili River. The general scheme of the system and the layout of the drainage network are presented in Figure 6.9.

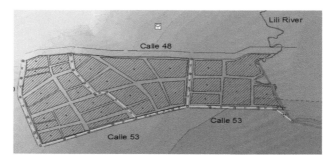

Figure 6.9 Scheme of the urban wastewater system for Sector A1 – Cali.

The design of the drainage is based on a pre-defined layout that follows the main roads proposed for the area to be urbanized. A combined drainage system was designed considering sewerage production and the runoff for a design rainfall event and the criteria defined in the preliminary studies carried out by the utility company (EMCALI and Hidro-Occidente SA 2006). The characteristics of the systems are presented in Table 6.2.

Table 6.2 Design characteristics for the drainage system and the wastewater treatment plant.

Sewer			CSO + Tank			WwTP		
Characteristic	Value	Unit	Characteristic	Value	Unit	Characteristic	Value	Unit
Area	85	ha	Setting	420	l/s	Design capacity	49250	PE
Length	3.8	km	Storage	2604	m^3	Solid retention time	12.7	d
Population	25000	PE	Overflow	8480	l/s	Anoxic reactor	2000	m^3
Return Period	2	years				Aerated reactor	7125	m^3
Rainfall	91.5	mm/hr				Secondary settler	1728	m^3
DWF	75	l/s						
WWF	8900	l/s						

DWF: dry weather flow, WWF: wet weather flow and PE: population equivalent

The system was modelled using EPA Storm Water Management Model (SWMM); the details of the modelling are presented in Chapter 4. The case considered for the experiments includes 25 pipes that form the main sewer network, the setting of a CSO and the flow pumped to a WwTP. These variables are used to optimize two objectives: flood damage in the urban catchment and the cost of the UWwS. The variables are constrained to fulfil engineering criteria.

The decision variables X were defined as follows:
- From the 25 pipes in the system only the last 11 pipes were optimized. This means, the size of pipes of the three branches shown in Figure 6.9 were fixed in the optimization process.

- Pipe sizes were selected within 10 possible options according to discrete sizes from the manufacturers. The boundaries of the pipe sizes were defined using as a mean size the estimated pipe in the preliminary design.
- The roughness coefficient was fixed assuming that only one type of pipe was used (concrete)
- The slope (s) was included in the algorithm as a constraint, such that s guarantees a minimum velocity for self-cleaning pipe and the cover depth is $>= 1m$.
- The storage capacity was included as one decision variable (volume). The boundaries for storage volume were fixed based on the preliminary design of the system.

Despite the simplifications in the case study, the search space of solutions is huge. Assuming the storage volume as a discrete variable with 10 steps, the possible combination of solutions to the problem is 10^{12}. Therefore the problem requires an important number of runs to approximate the true Pareto solutions. The results of the experiments are presented in what follows.

6.6 Results of Optimization Using Parallelization and Cloud Computing

6.6.1 Evaluation of instances with sequential optimization algorithm

Amazon EC2 offers four different types of instances named as small, medium, large and extra large; depending on the characteristics of the machine. The first experiment was the evaluation of the performance of those instances associated with the UWwS optimization using a sequential NSGAII algorithm. The benchmark model was set up in such a way that in each instance the objective function was evaluated 25 times in a single core. For comparison, a computer from the local network was also evaluated. Table 6.3 shows the characteristics of the instances available in Amazon EC2 and the desktop computer called "local". The results of the sequential optimization experiment are presented in Figure 6.10 and Table 6.4.

Table 6.3 Characteristics of Amazon EC2 instances and the local computer

Instance Type \ Features	Small	Medium	Large	Extra Large	Local
Number of cores	1	2	2	8	2
Memory (GB)	1.7	1.7	7.5	68.4	2
Storage (GB)	160	350	850	1690	300
Platform (bit)	32	32	64	64	32
I/O Performance	Moderate	Moderate	High	High	High
Price ($/hour)[1]	0.12	0.29	0.48	2.88	-
EC2 Compute Units/core[1]	1	2.5	2	3.25	3.52

[1]Price as of 2009. [2]One EC2 Compute Unit provides the equivalent CPU capacity of a 1.0-1.2 GHz 2007 Opteron or 2007 Xeon processor. Adapted from *www.amazon.com/ec2*.

The actual computing times for each type of computer are presented in Figure 6.10. The results correspond to the average computing time (T_s) estimated by running the same optimization experiment in sequential mode five times. The column representing the time was divided into two parts: the upper part corresponds to the computing time spent in 25 evaluations of the function (T_f) and the lower part is the rest of the computing time (T_q). T_q represents the time spent by the optimization algorithm doing the mutation, crossover and selection processes.

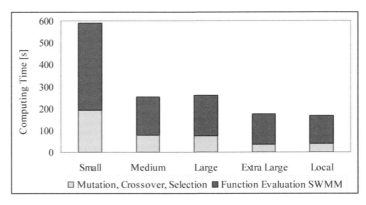

Figure 6.10 Computing time of a UWwS optimization using sequential algorithm and different computer characteristics

An important feature to highlight from Figure 6.10 is that the time spent in function evaluations is between 68% and 79% of the total time. Therefore, we can deduce that using the master-slave approach, a maximum of 80% of the optimization process can be parallelized; the rest can be regards as the non-parallelizable portion. Contrary to what it is assumed in the literature $T_s = T_f = nt_f$ (Cantu-Paz 1999); the non-parallelizable portion in the NSGAII implementation used here seems to contribute a percentage that may not be neglected. In other words:

$$T_s = T_q + T_f = T_q + nt_f$$ Eq. 6.2

Where:
T_s: computing time in sequential optimization mode
T_q: time non-parallelizable
T_f: computing time of all function evaluations
n: number of function evaluations
t_f: time of one function evaluation

The experiments were done using a single processor with a 32 bit encoding. Thus, number of cores and platform are fixed for the case study. The optimization process has low data requirements, and data is locally stored, therefore storage is not an issue for these experiments. Input/Output performance may be very important for communication in a cluster arrangement. Perhaps, the key feature for the optimization experiment is the EC2 compute units/core, in other words, the capacity of the processor. To compare the performance of the computers we used two indicators: the speed-up using as reference the time of type "Small", and the Speed-up/Price ratio. The results are presented in Table 6.4. The first two lines in Table 6.4 show a linear relationship between capacity of processor and Speed-up. That is, for a sequential optimization, the bigger the processor capacity the better the speed-up obtained. The shorter computing times were found with the local PC and the Extra Large instance. However, the speed-up/cost ratio shows that the best instance for the sequential optimization of the UWwS analyzed is the Medium type.

Table 6.4 Performance indicators of computers for sequential optimization experiments

Instance Type / Features	Small	Medium	Large	Extra Large	Local
EC2 Compute Units/core[1]	1	2.5	2	3.25	3.52
Speed-up	1	2.33	2.27	3.36	3.54
Speedup/(price/core) ratio	8.33	16.07	9.46	9.43	-

[1] One EC2 Compute Unit provides the equivalent CPU capacity of a 1.0-1.2 GHz 2007 Opteron or 2007 Xeon processor.

Perhaps, applications that made use of the multiple processors, 64 bit platform and demand storage capacity, may yield more from the Extra Large instances. However, for the cases-study of UWwS optimization that is not the case. Based on the results found in this experiment two types of instances were selected to form the virtual clusters: Small and Medium. The local cluster is also composed with PCs with similar characteristics to the one evaluated in this experiment.

6.6.2 Evaluation of different clusters with parallel optimization algorithm

Results of the experiments

The aim of the experiment is to evaluate the performance of different sizes of clusters with different types of parallel computer infrastructure, and to explore an optimal scale of the virtual cluster associated with the research problem. The optimization algorithm was set up with a population of 24 individuals and 20 generations. Three clusters were evaluated: two virtual, one formed with computers type small called a "small cluster" and other formed with medium type computers called a "medium cluster"; the third cluster was formed with computers from the local network of UNESCO-IHE and was called a "local cluster". The size of the clusters was variable, that is, the number of computers that formed the cluster was: 2, 3, 4, 6, 8 and 12. The master computer takes a portion of the individuals to evaluate the function, therefore it behaves also as slave. The population size and the scale of the cluster were selected in order to have an exact task-to-processor mapping. The idea was to reduce the inefficiency caused by processors that have to stay idle until they receive another task.

The results of the experiments with different clusters are presented in Figure 6.11. The computing time reported corresponds to the average of five runs of the same optimization process. The total computing time (Tp) for each size of cluster is plotted in Figure 6.11 (a). As was expected the small cluster has a slower computation time than the medium and local clusters. The trend of computing time follows a power function pattern. Both types of clusters (local and virtual) show similar patterns reaching an asymptotic curve for a cluster size of 12 computers. That means that a cluster with size 12 may be close to the maximum reduction feasible for the problem being studied.

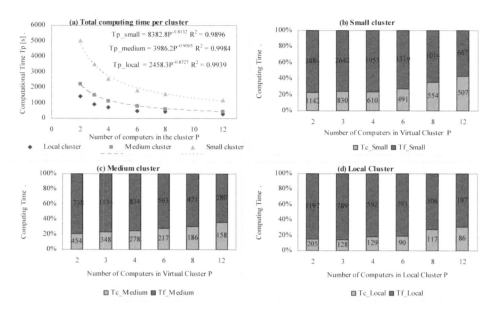

Figure 6.11 a) Total computing time per cluster and distribution of computing time for b) Small cluster, c) Medium cluster and d) Local cluster.

Figure 6.11 b to d, show the distribution of the computing time in two components. The upper corresponds to computing time spent in evaluations of the objective function (T_f). The lower part is related with the time spent in the sequential processes of the optimization algorithm and the communication between master and slaves ($T_q + T_c$). The three clusters follow the theoretical behaviour of a master-slave approach. That is, though computing time decreases as size of cluster increases, the T_c contribution to the total computing time increases. Comparing Figure 6.11 b and d, we can found that T_c percentage increase faster in Small clusters than in Local clusters. Therefore, we can deduce that for optimizations with fixed problem size, the faster the processor the more room for computational reduction.

Following the theory of master-slave approach the computing in a parallel optimization could be mathematically represented as:

$$T_p = T_c + T_f \qquad \text{Eq. 6.3}$$

Where: $T_f = n/P*t_f$
   n: number of function evaluations
   t_f: time of one function evaluation
   P: number of computers in the cluster
   Tc: communication time between master and slaves
   Tc = TSend + Treceive
   TSend : time to send the individuals to be evaluated
   Treceive: time to receive back the function values
   Tsend = S(Bn/p*li + L)
   Treceive = Bn/P*lf + L
   S :number of slaves = P-1
   B : inverse of bandwidth
   li : amount of data send
   lf: amount of data received
   L: latency

The communication time (T_c) is variable and depends on the amount of information transferred (l), the inverse of the bandwidth (B) and the latency of the communication network (L). Thus, $T_c = Bl + L$. Even though Equation 4 is very detailed, normally B is assumed to be too small to be considered; therefore only the latency is estimated. The communication time ended up being treated as a constant value times the number of slaves $T_c = S*L + L = P*L$, where L is a constant.

The network connection at UNESCO-IHE has a speed of 100 Mbps and a propagation time estimated as 5.0 µs. The size of the vector of decision variables that must be sent to the slaves is approximately 33 bytes and the objective function values received are around 13 bytes. Therefore, the communication time T_c for the Local cluster can be estimated as:

$$Tsend = 5 \text{ µs} + 33 \text{ bytes}*8bit/100Mbps = 7.64 \text{ µs} \qquad \text{Eq. 6.4}$$

$$Treceive = 5 \text{ µs} + 13 \text{ bytes}*8bit/100Mbps = 6.04 \text{ µs} \qquad \text{Eq. 6.5}$$

$$T_c = n*(Tsend + Treceive) = 480*(7.64 + 6.04) = 6.5 \text{ ms} \qquad \text{Eq. 6.6}$$

According to the estimation of T_c, it appears that communication and latency are not as significant as initially thought in the estimation of the computational time. It seems that, the basic sequential calculations done in the Master computer (crossover, mutation, selection, distribution), may be demanding most of the computing time considered for communication. Using the information collected in the experiments, we deduced the sequential and communication time required to process each individual in the population of the optimization process. The results of this estimation show that t_c is not constant; therefore the total time Tc is not constant. A good representation was found with a linear regression as illustrated in Figure 6.12

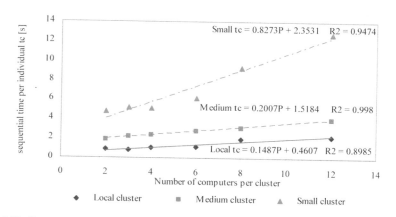

Figure 6.12 Computational time of sequential tasks per individual in the optimization process

Considering that the sequential time is not constant ($t_c = a*P + b$), the proposed mathematical representation of the total computing time is presented by Equation 6.7.

$$T_p = n*P^{-1}(a*P + b) + n*P^{-1}*t_f \qquad \text{Eq. 6.7}$$

With the information collected in these experiments is not possible to deduce the reasons of the increasing trend of computing time with the increment of the cluster size. With the availability of high speed networks, the communication time must be re-evaluated.

Performance indicators of parallel optimization
In order to evaluate the performance of the parallelization, three indicators were estimated: speed-up, efficiency and time reduction. Speed-up is estimated as the fraction of execution time in sequential mode (T_s) to the execution time in parallel mode (T_p). The increasing trend of the speed-up indicator is shown in Figure 6.13. The proximity of Local clusters to the ideal speed-up implies a better scalability for clusters with more powerful processors. However, in all three types of clusters when the number of instances increases, the sequential computation and communication time (T_c) rises and gradually reduces the speed-up of the clusters.

The Master-Slave approach is more efficient for problems where T_f is much bigger than T_c ($T_f \gg T_c$) (Cantu-Paz 1999) . In the hypothetical case study implemented here, T_f is between 5 and 3 times T_c . Therefore, T_c approaches quickly T_f as the number of computers in the cluster increased. In practical applications, the problem size of an urban drainage optimization will easily make T_f much bigger than T_c, because it requires more function evaluations with longer hydraulic computations.

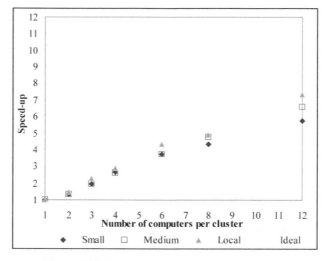

Figure 6.13 Speedup in different scales of cluster

Efficiency is the speed-up divided by the number of instances. The efficiency of computation is presented in Figure 6.14. For a perfectly parallelized program, the efficiency would be 1. Lower numbers indicate lack of full parallelization and/or the effect of serial and communications overheads (Creel 2005). In other words, there is no sense in continuing to increase the size of a cluster, if the efficiency is less than 50% of their potential (Cantu-Paz 1999). In both, virtual and local clusters, the declining trend of the efficiency is similar. The main difference is that for slow processors (Small clusters) the efficiency declines faster for clusters sized 6 and 12. Small clusters seems to reach the maximum speed-up when 12 computers are used; the use of more computers will make the optimization inefficient.

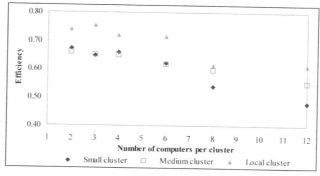

Figure 6.14. Decreasing trend of efficiency of parallel computing

Beside the good performance of the Local cluster, it has a stability issue as found by other researchers who use the idle capacity available in networks of work stations to form the cluster. When a user of a PC returns to the workstation unexpectedly and starts using it, the performance of the local cluster drops. That is what happen during the experiment with the Local Cluster with size 8 (Figure 6.14, "local cluster"). Even though the work station was automatically replaced by another available in the network, the performance was affected. In contrast virtual clusters have shown no sign of instability during the experiments.

The reduction of computing time for each type of cluster is shown in Figure 6.15. The maximum improvement reached was 86% with the Local cluster with 12 computers. In general, for the case study, the optimal size of cluster seems to be around 8 computers, for which the computational reduction is approximately 80%.

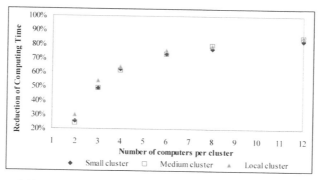

Figure 6.15. Improvement of performances versus different cluster sizes

In summary, parallelization of NSGAII seems to be a good alternative to reduce computing time. For practical applications, the Master-Slave approach will be even more efficient than what has been achieved. In terms of parallel infrastructure, cloud computing seems to perform at the same level of the Local cluster. Even more, characteristics like homogeneity of computers, facility to increase the size of the cluster and low maintenance requirements are key advantages when compare with local clusters. The use of fast communications networks seems to produce an important reduction in the effect of communications on the total computation time. A cluster with 8 computers seems to be affordable for engineering companies and practitioners, making this alternative a real option to reduce the total computing time. The reduction of 80% of the computing time is already a challenge for the next approach tested, the use surrogate modelling, which is described in what follows.

6.7 Results of Optimization Using Surrogate Models

6.7.1 Optimization using process-based model

To establish the base-line of the performance indicators, a standard optimization using the sequential NSGAII algorithm and the SWMM model of the UWwS (mechanistic model) was executed. The results of the optimization are presented in Figure 6.16. Two objective functions were minimized: the Total Cost and the Flood Volume. The sets of Paretos illustrate the evolution of the non-dominated solutions throughout different generations. Because the problem is a minimization, the Pareto set moves from the expensive part of the cost function space (generation 1) towards the cheapest part of the cost function space (generation 19). Because the objectives are contradictory, the cheapest solution found corresponds to higher flood volume. Another important characteristic of the evolution is that the Pareto set found in generation 1 contains only a few solutions while generation 19 has a denser and more uniformly distributed set of solutions.

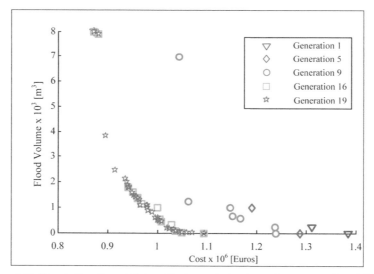

Figure 6.16 Pareto frontiers obtain with the optimization using process-based Model

The performance indicators, hypervolume and spread where estimated for the Pareto sets found in each generation. The values of the indicators are plotted in Figure 6.17. In this step the indicators are used to assess the quality of a Pareto solution set as obtained by the standard NSGAII algorithm. The values of hypervolume increase exponentially until they reach a stable Pareto set of solutions (Figure 6.17 a). This means that the Paretos dominate a wider area in the solution space as the generations evolve. The differences in the hypervolume indicators for generations 18 to 20 are very small; therefore we could establish that with the standard optimization, 19 generations are required to find the best Pareto front with a index HV = 0.666.

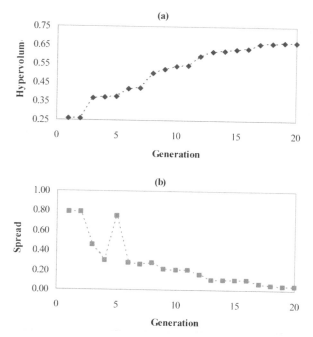

Figure 6.17: Performance indicators for Pareto sets obtain with the optimization using process-based model: a) Hypervolume and b) Pareto spread.

The Pareto spread indicator shows a decreasing exponential trend (Figure 6.17 b). This means that as the algorithm evolves, the solutions found are better distributed in the Pareto frontier. The spread of values for generations 19 and 20 is the same. Thus, this indicator confirms that 19 generations are required to obtain the best Pareto front. The Pareto spread figure shows a jump at generation 5, which means that there is a deterioration in the distribution of solutions with respect to the Paretos found in generations 3 and 4. The indicator seems to be very sensitive to the number of solutions in the Pareto set. The loss of one solution between generation 4 and 5 may have caused the increment in the average crowding distance. Despite the weak point found in the indicator, the trend and the final index S = 0.039 are suitable as a base-line for comparison.

The number of evaluations in the mechanistic UWwS model can be estimated as the sum of the evaluation of an initial population plus the generations required times the elements in each population. Therefore the base line of the computing time indicator for the standard optimization is:

$$Fe_m = 100 + 19*100 = 2000 \text{ (evaluations in the mechanistic UWwS model)}$$

6.7.2 Optimization using surrogate modelling approach

Analysis of learning algorithms

Before implementing the surrogate methodology a series of analyses were carried out to assess the performance of the learning algorithms selected. The optimization problem include

173

two criteria: cost and flood volume. The first criterion is a power function that does not need to be approximated. The second function, flood volume is the one that is computationally demanding. Therefore, the data for training the algorithms was created by applying a random sampling of the variables and estimating the corresponding flood volume with the process model for the UWwS. The algorithms were trained on 30 data sets producing four surrogate models for each set. Each model was validated using a new series of data. The performance of each surrogate model was assessed using two indicators: Root Mean Square Error (RMSE) and coefficient of determination (R2). The results of the test are presented in Figure 6.18.

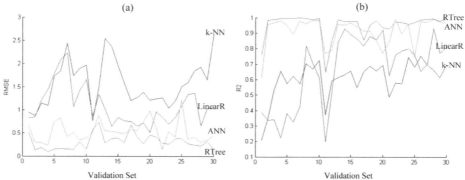

Figure 6.18 Performance of learning algorithms trained to approximate the flood function. a) Root mean square error b) Coefficient of determination

According to the performance indicators, the learning algorithms that better approximate the flood function are RTree, followed by ANN. As was expected the simpler structure of LinearR and k-NN has a less accurate approximation to the flood function (Figure 6.18 a). k-NN and LinearR also present the highest variability in the indicators (Figure 6.18 b). These two learning algorithms seem to be more sensitive when there is a need to explore different areas of the search space than the areas used for their training. Figure 6.19 summarizes the performance indicators estimated for the four learning algorithms. From the figure we can conclude that RTree and ANN represent the basic model better and have better predictability of the flood function than LinearR and k-NN algorithms. This may be associated with the fact that RTree and ANN are better equipped to represent complex non-lineal objective functions. However, one disadvantage that can be mentioned is that they may demand more computing time than a simpler function like LinearR. For testing procedures, the training time is not significant but for more complex applications this could became a factor to consider. Despite the low performance of the k-NN and LinearR, we continued the evaluation of the optimization with the four algorithms, considering that the methodology proposed here is less dependent on the accuracy of the surrogate model.

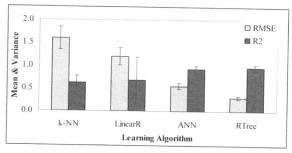

Figure 6.19 Mean and variance of the performance indicators of learning algorithms

Surrogate model optimization

Once the algorithms to be used for the surrogate model were selected, the methodology proposed for the optimization was carried out. The performance indicators of the ten iterations are presented in Figure 6.20. The hypervolume is presented in Figure 6.20 (a) and the Pareto Spread in Figure 6.20 (b). Each point on the curves represents the performance indicator measured for the final Pareto Set as found in the iteration. In general, the four surrogate models found solutions with similar indicators and they all seem to converge.

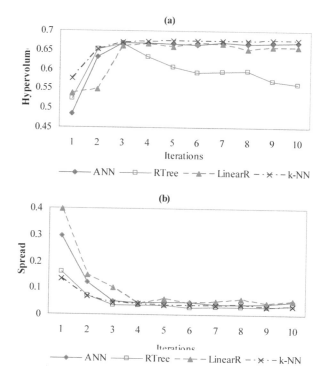

Figure 6.20: Performance indicators for Pareto sets obtain with optimization using surrogate models: a) Hypervolume and b) Pareto spread.

The hypervolume shows an increasing exponential trend stabilizing at the third iteration (Figure 6.20 a) for the optimization based on ANN and k-NN. The rising limb may appear very stiff, but it has to be considered that by the end of iteration 1 the optimization has already evolved 25 generations with the surrogate model. Therefore, the values of hypervolume are consistent with the values found for the standard optimization (Figure 6.17 a). Even though the surrogate based optimization process could have been stopped at iteration 3, the algorithm was left to finish the 10 iterations. An important feature appears after the third iteration, the optimizations based on ANN and k-NN continue to give relatively stable hypervolume values but the LinearR and RTree seem to lose performance. RTree especially shows significant degradation in the hypervolume indicator after the third iteration. This may be associated with the addition of irrelevant information to the samples used for training the algorithm. According to Michell (1997), Regression Tree algorithms show significant degradation in performance when irrelevant features are added. In practical terms, a decreasing hypervolume means that the solutions found by the optimizations based on RTree after the third iterations may contain sub-optimum solutions (i.e. solutions that may be dominated).

The Pareto spread shows a decreasing trend stabilizing at the third iteration for the optimizations based on ANN, RTree and k-NN, and at the fourth iteration for LinearR (Figure 6.20 b). All the optimization processes based on surrogate models were able to find wide spread solutions in the Pareto Front. Additionally, with exception of the LinearR, there is no deterioration in the values of the spread. This means that even for RTree for which Pareto Solutions goes(?) backward after the third iteration, the solutions were still uniformly distributed.

Considering both indicators the results show that the ANN and k-NN are the best surrogate models for the optimization process. Surprisingly, the simplest model k-NN seems to be the fastest and more stable surrogate model for the case study analyzed. The k-NN algorithm may have been favoured by the sampling process based on the genetic algorithm used for optimization. The samples used to train k-NN are Pareto solutions, therefore they are independent and widely distributed at the front, helping the training algorithm to converge and minimize the misclassification error.

Knowing that the methodology works properly and finds solutions with the same quality as the optimization based on the UWwS model, we focus in the main objective, the reduction of the computation time. To assess the time reduction we compare the number of function evaluations for all the optimizations. The results are shown in Figure 6.21; the upper figure compares hypervolumes and the lower figure the Pareto spreads. This figure presents the actual time reduction that benefit from the fast running surrogate models.

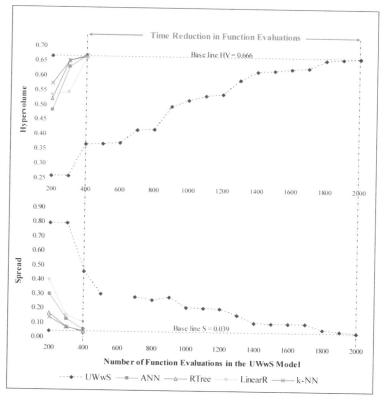

Figure 6.21 : Computing time reduction by using surrogate modelling approach.

As can be seen in Figure 6.21, to reach the base line, the optimization based on fast surrogate models requires 400 evaluations of the model for the UWwS. In contrast, the standard optimization requires 2000 evaluations to reach the same quality of Pareto solutions. Therefore, the speed-up can be computed as sp = 2000/400 = 5. In other words, surrogate optimization approach reduces the computing time by 80% for the Cali case study. These results are consistent with the values presented by Fu, et al (2008) when testing ParEGO versus NSGAII to optimize real time control strategies in UWwSs.

6.8   Discussion

Parallel multi-objective optimization
Parallel multi-objective optimization is an established approach that has been widely applied in computationally demanding problems. The master-slave approach appears to be very useful for the optimization of an urban wastewater system in which the function evaluations are time consuming. The results for the case study show that parallel multi-objective optimization can speed-up 5 times the computation when using the capacity available in a cluster efficiently with 8 computers. In other words, the computing time is reduced by 80%.

In order to maximise the benefits in computing time reduction, the design of the experiment should consider the following points:

- The population size of the MOEA must be evenly divisible between the slaves in the cluster so as to have an exact task-to-processor mapping. The idea is to reduce the inefficiency caused by processors that have to stay idle until they receive another task.

- The computing time of one function evaluation should be very similar throughout the whole optimization. Non-feasible design variables may create instabilities in the hydraulic model and increase the hydraulic computation. Therefore, modifications to the optimization algorithm should be carried out in order to guarantee that the vector of design variables has feasible solutions.

- The computing time for the function evaluations must be bigger than the time used by the algorithm in the sequential and communication part of the process ($T_f >>> T_c$). In practical applications of urban wastewater optimization, this should not be a constraint because the main reason for parallelization is the high computing demand of the hydrodynamic and water quality models.

- In practical applications the amount of data required for the design of UWwSs can be significant. Therefore, the data should be pre-installed in the slaves. Reducing the amount of the data transferred to slaves during the optimization process will minimize the communication time required.

- Parallel infrastructure with high speed processor will benefit the scalability of the cluster and therefore the efficiency in computing time reduction. Similarly, cluster infrastructure with high I/O performance (>1 Gbps Ethernet with low latency) will reduce the time demands for communication and will benefit the efficiency of the clusters.

Some of the limitations encountered with the master-slave approach are:

- The MOEA algorithm (NSGAII) is not fully parallelizable. Evolutionary operations like crossover, mutation and selection are done in sequential mode. The sequential part of the algorithm, together with the communication time between master and slaves, reduces the efficiency of the clusters. In practical applications this may not be an issue if the cluster is designed with an optimal size. In addition, for a UWwS, T_c should be much smaller than T_f, therefore the sequential part of the optimization will start limiting the efficiency of the optimization when the high reduction of the computing time has been reached (possibly above 80%).

- In practical applications, the exact mapping between the task and the processor may not always be feasible. If larger differences exist in the time needed to evaluate different design solutions, the loading of the slaves may be uneven. An option to overcome this problem is the use of dynamic computational load-balancing across the slaves.

- The current available mathematical description of the computing time in parallel may be based on assumptions that have to be re-evaluated. The mathematical description developed in the 90s should now consider new developments in network communication and processing.

One of the main results presented in this chapter is the comparison between two types of parallel computer infrastructure: clusters formed with networked workstations and clusters formed with virtual computers in the cloud. The indicators evaluated show that virtual clusters have the same performance trend as the local cluster of workstations. The differences in the indicators are mainly associated with the capacity of the processor used to form the cluster.

The main benefits of use clusters in the cloud are as follows:

- The number of processors in the cluster can be increased rapidly and relatively easy. The use of AMIs to create instances with same configuration and data facilitate the scaling up of the cluster.
- The possibility of creating homogeneous clusters benefits the balanced loading of the slaves.
- When compared with clusters based on networked workstations, the cluster in the cloud seems to be more stable and reliable. The reason is that once the cluster has been created, the computers are dedicated exclusively to the optimization process while workstations may be disturbed by other users.
- The communication network available seems to have a high performance with a low latency. This characteristic may reduce the effect of the communications in the total computational time of the optimization and therefore the efficiency of the cluster.
- The infrastructure as a service could enable scientists and engineers to use parallel infrastructure on demand. One of the main benefits is that virtual clusters do not require initial investment and maintenance such as required by a dedicated parallel infrastructure. Although the cluster based on workstations will benefit from existing computing capacity, there is a cost involved in building and maintaining such an infrastructure.

The prospects of cloud computing are promising. Amazon EC2 offers the possibility of using already created clusters with very high performance, all in terms of processors and in terms of resources available and bandwidth with low latency. Considering this, the performance of the parallelization depends on processor and communication capacity; it will be interesting to experiment with this type of cluster on practical designs of urban wastewater systems. Amazon EC2 clusters, formed with virtual computers with a capacity of 33.5 EC2 compute units and very high I/O performance (10 Gigabit Ethernet) should be tested with practical applications of multi-objective optimization for UWwSs. One additional factor that may facilitate the parallel computing process is that MathWorks is working with Amazon Web Services to provide flexibility in hosting their applications. Therefore in the future, the parallel algorithms implemented in MATLAB will be available for research and engineering applications as part of the configurations of virtual computers. This trend could also be followed by the hydrodynamic software providers, closing the software – hardware needs for parallel multi-objective optimization.

Surrogate modelling optimization
The proposed approach, namely the Multi-objective Optimization by PRogressive Improvement of Surrogate Model (MOPRISM), seems to have great potential to reduce computing time in the multi-objective optimization of urban wastewater systems. The results for the case study show that surrogate multi-objective optimization speeds up the computations 5 times. In other words, the computing time can be reduced by 80% using the fast approximation of objective functions obtained with surrogate models.

One of the main results of the experiments is the success of the proposed surrogate model approach. In early research with model-based optimization, surrogate models were trained once and then used instead of the process (e.g. hydraulic) model (Solomatine and Torres (1996), Maskey et al. (2000)) in model-based optimization. Based on further development of the powerful ideas of stepwise retraining and improving the surrogate model (Nain and Deb (2002), Knowles (2005), Liu, et al. (2008)); we have presented the MOPRISM approach in this chapter.

The MOPRISM approach is closer to the approach presented by Liu, *et al* (2008). The comparison of those two methods is illustrated in Figure 6.22. The Liu, *et al* method shown in Figure 6.22a employs an approximation model management technique to explore the design space and to identify the Pareto optimal set. The algorithm sequentially moves the limits in the design space and verifies the actual Pareto optimal set during the iteration procedure. The evaluated Pareto optimal solutions are stored in an external archive and the archive keeps being updated at each iteration step. In other words, the algorithm develops local surrogate models in the design space. On contrast, MOPRISM develops local surrogate models in the objective function space (Figure 6.22b). The MOPRISM algorithm uses the surrogate model to approximate the objective function directly (response surface), and not the process model itself. At each iteration the MOEA explores the whole design space using evolutionary operations. The approximate Pareto solutions are verified with the process based model and these are used to update the sample. Therefore, the samples in the MOPRISM approach progressively improve the surrogate models and generate local models in the function space.

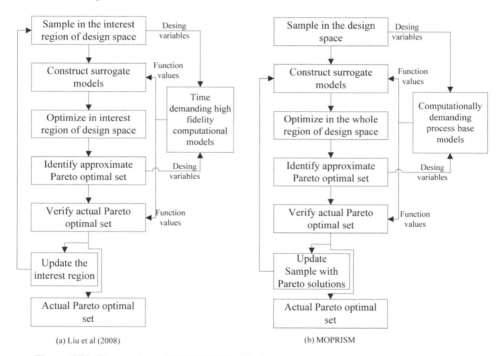

(a) Liu et al (2008)   (b) MOPRISM

Figure 6.22 Comparison of MOPRISM with the approach proposed by Liu, et al (2008).

The key issues of surrogate multi-objective optimization approach are as follows:
- The selection of the data-driven model to be used as surrogate model. Four different methods are tested and implemented as surrogate models (LinerR, ANN, k-NN and RTree). For the case study, ANN and RTree show similar results, RTree being the best approximator according to the results with the validation data set. As expected, the simple structure of Linear R and k-NN makes these two models less accurate. However, for the surrogate optimization process, the differences in accuracy shown in the initial test were not reflected. RTree was the worst surrogate model with a quick deterioration

of the performance in the optimization after the third iteration. It seems that the MOPRIS approach is less dependent on the accuracy of the model as also is claimed by Liu, *et al.* (2008) in their approach. More research is needed here to identify the reasons for the apparent deterioration of the performance of RTree.

- The sampling of the variable space. MOPRISM use in the initialization a random set but the sample is updated after the first iteration with the set of Pareto non-dominated solutions. The evolutionary operations of NSGAII are used to update the sample. It is important to highlight that for the urban drainage optimization used as a case study, the solutions that form the Pareto in the function space may occupy very different places in the design space. In other words, for urban drainage optimization, it may be beneficial to maintain a wider design space, contrary to what is proposed by Liu, et al (2008), which seems to reduce the region of interest. Further research with MOPRISM could include the evaluation of the behaviour of the design space as the optimization progresses. This could be done for instance by using the hypervolume indicator.

- The number of samples. This is critical because if the number is too small the data is insufficient to create a surrogate model, but if the sample is big the evaluation in the process model may limit the efficiency of the method in reducing the computation time of the optimization process. For the case study we used 100 individuals in the population of NSGAII[this may have hindered the speed-up obtained. Therefore, more research is needed to define the proper size of the sample.

- Converging criteria. Even though for the experiments we used as a stopping criteria a fixed number of iterations (10), the proposed criteria for future applications are hypervolume and average crowding distance.

- The reduction of time. Two criteria were used to estimate time reduction: the speed-up and the percentage of reduction. The speed-up factor of 5 is in the range reported by Liu et al (2008) for numerical test functions. Using μMOGA as the optimization algorithm and the processes described in Figure 6.22a they found a speed-up of 10 for the simple numerical test and a speed up of 28 for a simple engineering problem. For more complex problems they did not report any speed-up. Further research is needed to compare both approaches using numerical test functions and benchmark optimization problems.

Computing time reduction with parallel and surrogate optimization
Due to the differences in the approach and the different indicators used, the comparison here has methodological limitations. However, from a theoretical point of view, it is interesting to compare results. Table 6.5 presents a qualitative comparison of the two approaches presented in this chapter. The criteria and the values are based on the experience of the author. In general, we could say that both approaches can achieve a similar computing time reduction. The main difference is in the possibility of controlling the performance of the algorithm. In surrogate models, additional convergence criteria have to be used to guarantee a successful achievement of the Pareto set of solutions, while in the Parallel computing using the Master-Slave approach, the MOEA itself takes care of convergence. In Parallel computing you can design the experiment in order to have a certain time reduction while in surrogate modelling it is less controllable. However, parallel computing will demand an additional infrastructure and that may not be available, or the resources for the use of cloud may not be available. Therefore the implementation of parallel computing could be limited while surrogate modelling can be implemented on any available machine.

Table 6.5 Theoretical comparison of multi-objective optimization using parallel computing and surrogate modelling

Criteria	Multi-objective optimization	
	Surrogate modelling	Parallel Computing
Require special software	Data-driven model	Distributed computing
Require especial computer infrastructure	No	Yes: Parallel computing infrastructure
Demand special knowledge and skills	Data-driven modelling	Distributed computing and networks
Require addiitonal coding	Yes	Yes
Implementation time	Short	Medium to long
Cost of implementation	Low	Medium to High
Computing time reduction	Medium to High	Medium to High
Control of performance	Low	High

Further research could include a combination of the two approaches: an algorithm that combines the benefits of the surrogate model optimization and that at the same time is parallelized and run in a parallel computing infrastructure.

6.9   Conclusion

Conclusions of Parallel Computing on the Cloud
Based on the conducted experiments it can be concluded that Cloud computing can facilitate the building on demand of a parallel infrastructure needed for the optimization of a UWwS. The speed-up found in the virtual clusters was similar to that for the local cluster and to those reported in literature for the Master- Slave approach. Performance of the virtual clusters in the cloud was comparable with the local cluster comprised of workstations in a local network. Additionally, no instabilities were experienced during the experiments, so the reliability of virtual clusters was even better than for the local cluster where the speed-up was affected by the migration of jobs when machines were no longer idle. In general, the Infrastructure as a Service provided by Cloud computing is definitely an enabling technology for researchers and engineers in that it requires parallelization to solve complex optimization problems of water systems.

Conclusions of Surrogate Modelling
In general, the results with surrogate optimization show a significant reduction in computing time. The orders of magnitude of the indicators for speed-up and time reductions are similar to those reached with an efficient size of cluster in the parallel optimization. The proposed approach, namely the Multi-objective Optimization by PRogressive Improvement of Surrogate Model (MOPRISM), seems to have great potential in reducing computation time in the multi-objective optimization of urban wastewater systems.   However, this is still exploratory research; therefore much more experimental work needs to be done in order to tune the key issues of the algorithm.

General conclusion
The experiments in this chapter show that there are significant reductions in computing time using surrogate optimization or parallel computing. Two novelty approaches are proposed: the use of virtual clusters in the cloud as a parallel infrastructure and the MOPRISM algorithm to implement surrogate multi-objective optimization. These two approaches may facilitate engineers in the use of MOEAs for model based design and control of an urban wastewater infrastructure. The advances in surrogate optimization and parallel computing are promising, and the benefits are so important that it is possible to suggest that in the near future all multi-objective optimizations will include one of those two approaches and may even be a combination of them.

7 Conclusions and Recommendations

7.1 Conclusions

The optimum design of urban wastewater components has been the dream of many researchers in the past three decades. The needs of wastewater managers to fulfil higher levels of efficiency in terms of the reduction of pollution impacts have increased the demand for such optimum design and control of the wastewater system. However, most of the efforts have been dedicated to optimizing the design of one component at a time considering one or a maximum of two objectives. In this thesis, the design has been expanded to include the dynamic interaction between components: sewer network, wastewater treatment plant and receiving system, and the minimization of pollution impacts.

The aim of the research is to understand the possible benefits and drawbacks of a design approach that includes the dynamic interaction between the wastewater system components and the adoption of multiple objectives when evaluating design alternatives. These objectives were addressed in three steps: the first was to develop a general approach here called the Model Based Design and Control (MoDeCo) approach. The second step was to assemble an integrated modelling tool linked to a multi-objective optimization algorithm. And the third step was to apply the approach to two urban wastewater system designs: the hydraulic design of a sewer network for Sector 1A in Cali – Colombia, and the functional design of the wastewater treatment of Gouda – The Netherlands.

MoDeCo approach can be described as a combination of the iterative design and model predictive design approaches as defined by Harremoës and Rauch (1999). Thus, MoDeCo starts with a pre-design that is based on traditional approaches and empirical rules of operation. The pre-design is used to build the model of the system, together with information from the other components of the system. Then, the model plays the role of representing the real world and is used to predict the performance of the system. Alternative solutions are automatically generated and a set of optimum solutions is found using Multi-Objective Evolutionary Algorithms. The general conclusion is that solutions found with MoDeCo approach may have better performance than the design alternative using traditional methods. The main benefits and limitations of the approach are presented in what follows.

7.1.1 Benefits of the design using MoDeCo approach

- The design includes the interaction with other components of the UWwS. This allows designers to expand the scope of a sewer network design to include pollution impacts on the receiving system. The results of the case study for Cali - Colombia show that the optimum design of a sewer network considering only flooding and cost objectives, performs very poorly in terms of pollution impacts on the river (Minimum DO = 1.8 mg/l). In contrast, by including the water quality objectives, an optimum size of storage was found. Thus, the negative impact on the river was reduced to the minimum standard (Minimum DO = 4.1 mg/l) and at the same time, no flooding was generated in the urban catchment.

183

- The use of state-of-the-art modelling tools allows the design to be based on dynamic conditions contrary to traditional approaches that are based on constant design flow and water quality composition. The current functional design of Gouda is based on fixed rules selected for steady state conditions. However, the results of the experiments with the MoDeCo approach indicate that a functional design based on fixed set points is not the best alternative to deal with disturbances caused by precipitation events. The set points are dependent on two highly dynamic factors: the current status of the treatment plant, and the outflow and corresponding water quality composition from the sewer network. Therefore, it is possible to find optimum set points for each specific operational situation. For instance, in winter the effluent concentration of total nitrogen and total phosphorous were reduced by 51% and 53% respectively by optimizing the set points of the internal recycle flow and the dissolved oxygen concentrations in the aerated reactors. Even more, for highly dynamic sewer flows and water quality composition, the operation of the wastewater treatment plant may benefit by the use of anticipatory control, in other words, by adjusting operational variables in response to the dynamic interaction with the other UWwS components.
- Designs based on traditional approaches normally are sub-optimums. The search space of the design variables is so wide that it would be almost impossible to evaluate the alternatives in an exhaustive search and impossible to evaluate it in a laboratory scale model or the real UWwS. In MoDeCo approach, the use of optimization algorithms increases the chance of finding optimum solutions that consider multiple objectives.
- The design is driven by the minimization of cost while maintaining the performance of the system for other objectives. Traditional methods tend to use a safety factor that is normally reflected in oversized structures. Minimization of cost is therefore frequently one of the objectives of the optimization of the sewer network design. The results of the case study show that the optimization procedure can find solutions that are around 15% less costly than the sewer network designed using the traditional method of sizing pipes. This cost saving is comparable with the 5% to 30% reduction reported in the literature for the optimization of sewer networks (Guo, *et al.* 2008). In other words, the designs found by MoDeCo are more effective, because they may keep the same level of protection against flooding using fewer resources.
- The approach enhances the analysis of a great number of alternatives and the Pareto set of solutions gives a variety of design alternatives to be analyzed further by decision makers. In traditional approaches only a few alternatives are assessed. This is in contrast to the design of the sewer network for Cali, where up to 50000 alternatives were assessed. The final Pareto set of solutions in that case contained 153 solutions, which gives the possibility to analyze further trade-offs between the objectives according to the preferences of the decision makers.
- The designer and decision makers are better informed of the solutions and their consequences. In fact, perhaps the main benefit of the method is the information generated for each alternative solution. For instance in the functional design of Gouda WwTP, the ratio of internal recycle flow to the influent flow (Qir/Qin) tends to indicate that the optimum set points are more dependent on the flow conditions than on the temperature conditions, contradicting the current operational design. From the practical point of view this implies that a set of ratios for different influent flows, as the ones defined in Chapter 5, may help the operators refine the internal recycle setting.
- MoDeCo approach is in line with new regulations that enforce a holistic view of the urban wastewater management and the reduction of impacts on the receiving system (e.g. Water Framework Directive).

7.1.2 Limitations of the design using MoDeCo approach

- There is a lack of information to build the integrated model. This is an issue that is not only inherent to MoDeCo; in general, it is a limitation for any kind of methodology that tries to address in an integrated way the design of an urban wastewater system component.
- The complexity of building an integrated model demands different expertise and skills. In fact, this is not a job that should be done by one person but by a multi-disciplinary team.
- The optimum design depends on the accuracy of the model's predictions. And the uncertainty in the model may threaten the validity of the optimization process. Moreover, the success of the approach relies heavily on the skills, experience and judgement of the engineer that sets up and runs the models.
- The approach has a high computational demand. Two factors influence the computing demand. The first is that an integrated model of the system is computationally demanding in itself and second, the optimization process requires a significant number of function evaluations to converge to a set of optimum solutions. This may threaten the use of the method in a practical application.
- The design of an UWwS is a multi-variable and multi-objectives problem. But the more variables and the more objectives included the less probable that an optimum solution can be found. Thus a proper level of complexity must be decided, and a proper experimental design has to be prepared.
- As the final result of the process is a set of solutions, these solutions may require further analysis by experts to facilitate the decision making. The final selection of one design may combine criteria based on the non-modelled preferences of decision makers and performance indicators for resilience based on long term simulations.
- The integrated wastewater management requires interdisciplinary contributions and the will of the institutions in charge of system. The transfer of knowledge in this field may help to reduce possible resistance of design engineers to the use integrated models and optimization algorithms.

7.1.3 Reduction of computing time

One of the limitations was the long computing time required to find optimum solutions. A step forward to solve this problem was done by testing two alternatives to reduce the computing demand: parallel computing and surrogate modelling. In this research two novel approaches are proposed: the use of virtual clusters of computers in the Cloud as parallel infrastructure and a new surrogate modelling method here named the Multi-objective Optimization by PRogressive Improvement of Surrogate Model (MOPRISM).

Parallel Multi-Objective Optimization
Parallel multi-objective optimization is an established approach that has been widely applied in computationally demanding problems. The master-slave approach appears to be very useful for the optimization of urban wastewater system in which the function evaluations are time consuming. The results for the case study show that parallel multi-objective optimization can speed-up the computation 5 times using efficiently the capacity available in a cluster with 8 computers. In other words, the computing time was reduced by 80%.

185

Parallel Computing on the Cloud

Based on the conducted experiments it can be concluded that Cloud computing can facilitate the use on demand of the parallel computing infrastructure needed for the optimization of an UWwS. Performance of the virtual clusters in the Cloud was comparable with the local cluster comprised of workstations in a local network. The speed-up found in the virtual clusters was similar to the local cluster and to those reported in literature for the Master- Slave approach. Even thought, the reliability of virtual clusters was better than the local cluster because in some of the experiments the speed-up in the local cluster was affected by migration of jobs when machines were no longer available. In general, the Infrastructure as a Service provided by Cloud computing is definitely an enabling technology for researchers and engineers who require parallelization to solve complex optimization problems of water systems.

Surrogate Modelling

In surrogate modelling the computationally demanding process-base models are replaced by data driven models. In general, the results with surrogate optimization show a significant reduction in computing time. The results for the case study show that surrogate multi-objective optimization speed-up the computation 5 times. In other words, the computing time was reduced by 80% using the fast approximation of objective functions obtain with surrogate models. One of the limitations of surrogate models is that the accuracy of the data driven model may be reduced as the optimization algorithm explore different areas of the variable space. The proposed approach, Multi-objective Optimization by PRogressive Improvement of Surrogate Model (MOPRISM) addresses this problem by automatically re-training the surrogate model in a loop as the optimization algorithm progress towards the optimal set of solutions. Even though the promising results, more experimental work needs to be done to tune the key issues of MOPRISM approach.

Comparison between parallel computing and surrogate modelling

Overall, the experiments presented in this thesis show that there are significant reductions in computing time using surrogate optimization or parallel computing. In both cases, reductions up to 80% of the initial computing time were achieved. The main difference between the two alternatives is in the possibility to control the performance of the algorithm. In surrogate models, additional converging criteria have to be used to guarantee a successful achievement of the Pareto set of solutions; while in Parallel computing using the Master-Slave approach, the Multi-Objective Evolutionary Algorithm (MOEA) itself takes care of convergence. In Parallel computing you can design the experiment in order to have a certain time reduction while in surrogate modelling the time reduction is less controllable. Parallel computing demands additional infrastructure that may be supply by a local cluster or by using the Cloud. In surrogate optimization there is no demand for additional infrastructure but there is a need for knowledge and skills to set up data driven models.

These two approaches may facilitate the use of MOEAs by engineers for model based design and control of urban wastewater infrastructure. The advance in surrogate optimization and parallel computing are promising, and the benefits are so important that it is possible to anticipate that in the near future all multi-objective optimizations will include one of these two approaches, or may even be a combination of them. Finally, in the optimization of UWwS, the limitation of computing demand should be narrowed to exercises where serial MOEAs are used, and perhaps for on-line optimization of control strategies where the time available to find a solution is critically short.

7.2   Recommendations

7.2.1   Recommendations for practical applications

The design of an UWwS should be done considering the dynamic interaction between components. *There is no UWwS that has the advantage of operating with a constant flow rate and a constant water quality composition.*

UWwS have a life of their own. There are no two systems that are the same so you cannot generalize but only customize the solution for each system. In consequence, the design of each UWwS is a separate exercise.

Do not code all of your dreams because you may end-up having a nightmare analyzing the results. When this research started, the available software to do integrated modelling of urban wastewater system had a number of limitations. Among others, the better software modelling products were license protected so no changes could be made to them as needed for this research, and most if not all of the products were strong in one component but over simplified in modelling the other components of an UWwS. Thus, the decision was taken to couple existing open source state-of-the-art models for each component. But the result of this was a period of intense coding. By the time this thesis was being written, some of the integrated UWwS software products have been improved. Therefore a practical suggestion is to search among the available software products before starting to code your own. Further research could include the upgrading of existing public domain Integrated UWwS models.

Find a compromise between complex process-based models and simplified surrogate data-based models. In this research I tried to find and use the appropriate complexity of model to represent the components of the UWwS. Looking for that right level of complexity I chose to integrate state-of-the art "Process Based Models". However, when trying to optimize the system using the integrated model in pursuit of a degree of precision, the computing demand limited the ultimate goal of finding an optimal solution. Contrary to the initial choice, one of the best solutions to the problem was to replace the computationally demanding model with a more simplified surrogate model based on data-driven approaches.

Find a compromise between the time series of precipitation that generates the critical disturbance but does not limit the optimization process. In UWwS, one of the main disturbances is precipitation. There is no doubt that a time series of dry and wet weather periods should be used to assess the performance of any competing design alternative. However, to evaluate the performance of the system, the time series of precipitation may require long computing times that make the optimization process impractical.

The optimum cost of an UWwS is not the lowest cost. The minimization of cost seems to be one of the main arguments to justify the optimization of the design of an UWwS component. Here, I was also caught in a trap, hoping to find the ideal solution that will greatly reduce the cost when compared with solutions developed with traditional approaches. However, both case studies show that cost savings are modest. In each case, namely the optimum sewer network design for Sector 1A- Cali and in the optimum operation of the wastewater treatment plant of Gouda, the best solutions found cost more than the defined base scenario. Thus, the optimization should be driven by other higher level objectives such as: the protection of the community that is served by the UWwS and the protection of the ecosystem that serves the UWwS. The question remains: is the social and environmental consciousness strong enough to find supporters to embark in such a major undertaking knowing that the cost saving may at best be "modest"?

The design and operation of an UWwS considering the interaction with other components seems to be a promising alternative to achieve more sustainable development of urban areas. However, in practice each of these systems is managed separately. In Gouda for example, four institutions are involved in the management and operation of the components of the UWwS. In Cali, the sewer network and the treatment plant are managed by one utility company that is subdivided into units, each of which seems to work separately. In addition, the receiving systems are managed and controlled by two environmental authorities. Thus, implementation of an optimum design using the MoDeCo approach may be hindered if coordination between the relevant institutions cannot be achieved.

7.2.2   Recommendations for further research

Design and control of UWwS components
The sewer network design:
- In this research we have expanded the scope of the sewer network design. The inclusion of interactions with other components, the minimization of the pollution impacts and even the operational variables is a step forward towards an integrated design. Further integrated designs could include the optimization of the layout or the inclusion of ecological objectives in the receiving system.
- The results presented in this research are exploratory. More research is needed to demonstrate the effectiveness of integrated designs in other cases. Further research could be to define long term objectives, for example indicators of sustainability, resilience or robustness, and use them to evaluate the optimum solutions found in a post-processing step of the design.
- Even though effort was made to quantify accurately the initial investment cost of the sewer network being analyzed, more research is needed to complete the analysis including operational cost and replacement of infrastructure in a long term simulation.
- Most research into the optimum design of sewer networks uses combined or separated networks with circular pipes. This research was not the exception, even though the case was not trivial, a step forward could be the inclusion of sustainable urban drainage concepts (e.g. the optimization could include infiltration devices and local storages).

The wastewater treatment plant design:
- In this research we optimized the functional design of the wastewater treatment plant of Gouda. Further research should include a combination of the design of the wastewater treatment process in combination with the functional design.
- Research should also include control variables that influence slow processes like the growth of micro-organisms. A full functional design should also include state variables within the processes and longer operational horizons, for instance, including a period that allows the evaluation of changes in the sludge retention time or sludge production (i.e. two or three months).
- Anticipatory control of the wastewater treatment seems to improve the capacity of the system to deal with external disturbances (i.e. fluctuations in the influent wastewater quantity and quality). Research should be focused in the verification of the prediction of the disturbances, considering the propagation of the effect of the precipitation through the treatment plant. In addition, utility aspects like the cost saving generated by the use of anticipatory control strategies should be performed.

Integrated model of the urban wastewater system

The integrated model developed for this research was based on a combination of existing models of each subsystem. However, this approach has limitations because of: different state variables, a limited flux of information between models of the subsystems, and long computing times. Despite the efforts of developers of commercial software products that have incorporated for instance SWMM into the alternatives to model the sewer network, there is not yet a product that is integrated at all levels. Therefore, the development of a truly integrated model is still a task for researchers and software engineers. An alternative to reduce computing time is the development of Parallelized algorithms. Burgess, *et al* (1999) modified the FORTRAN source code of the EXTRAN block of SWMM4, which enabled the model to take advantage of parallel processors for faster execution during runtime. An interesting research project could be to parallelize EPA SWMM5 to enhance the use of readily available Duo Core and Quad Core computers.

Modelling water quality in the sewer network seems to be one of the weak points of the integrated model. Fundamental research is needed to understand and model better the build-up and wash-off of pollution on the surface of the catchment. Suggestions to improve the water quality modelling approach in SWMM are given by Sutherland and Jelen (2003). In addition, the sedimentation and re-suspension of cohesive material should be better understood and incorporated into the sewer model.

A proper calibration and validation of the models based on comparison with measured data should be performed. Further more, resources should be put to reduce the uncertainty in the models. The uncertainty in the result from the integrated model should be quantified along with a description of the effects of the uncertainty on the optimized solutions. The uncertainty should be propagated through the models of each component and a method to present the results of the optimization to the decision makers including the uncertainty should be developed.

Multi-objective optimization

In this research, the most common algorithm in urban wastewater optimization (NSGAII) was used as the optimization algorithm. However, further research could include other MOEAs and compare the results considering the criteria of convergence and computing demand. For instance, SPEA2 (Zitzler, *et al*. 2001) seems to find better distributed Pareto solutions for high dimensional objective optimization process. Micro-GA (Coello Coello, *et al*. 2001) seems to produce a Pareto set with a low computational cost, which might benefit the UWwS optimization. The use of hybrid algorithms (CA-GASiNO) that use a local search algorithm (Cellular Automata) to seed the NSGAII with well defined preliminary solutions and thus speed up the convergence to an optimal Pareto set (Guo, *et al*. 2007) should also be considered.

In the design of sewer networks, inequality constraints may complicate the topology of the search space, and optimization algorithms may generate unfeasible solutions. Alternatives to handle the inequality constraints should be explored further. In this research, the modifications of the operators in the NSGAII show potential for improving the final Pareto set when compared with the traditional method of penalty functions. Further research could include a customized reparation algorithm that takes unfeasible solutions from the engineer point of view and repairs them to the nearest known feasible solution.

Of great importance for practical applications of MoDeCo is the reduction in the computing time required to find the set of optimal solutions. The findings of the two alternatives tested in this research provide the following insights for future research:

From parallel computing approach
- The master-slave approach seems to be a straight forward approach for the parallelization of MOEAs. However, in the complicated topology of the search space of an UWwS optimization, the use of more efficient parallelization algorithms like the island model or the cone separation (Branke, *et al.* 2004) should be explored and compared with the commonly used master-slave approach.
- In terms of parallel infrastructure, Cloud computing seems to be a promising alternative. Amazon EC2 offers the possibility to use already created virtual clusters with very high performance. These clusters are formed with virtual computers with capacity of 33.5 EC2 compute units (i.e 40.2 GHz) and very high I/O performance (i.e. 10 Gigabit Ethernet). This seems to be a platform for parallel computing that should be exploited with practical applications of multi-objective optimization for UWwSs.
- The current available mathematical description of the computing time in parallel processing (Cantu-Paz 1999) seems to be based on assumptions that have to be re-evaluated. The mathematical description should now consider new developments in network communication and processing; for instance, the work of Gustafson (1988) on re-evaluation of Amdahl's law could be followed up.

From the surrogate approach
- The MOPRISM approach proposed in this research seems to have great potential; but further research is needed to clarify the method. For instance, the dependency of the results from the accuracy of the data-driven model used as surrogate model should be further explored. In addition, rules to select the number of samples used to create the surrogate model should be defined together with the stopping criteria for convergence of the optimization.
- For urban drainage optimization, it may be beneficial to maintain a wider design space, contrary to what is proposed by Liu, et al (2008), which seems to reduce the interest region. Further research with MOPRISM could include the evaluation of the behaviour of the design space as the optimization progress. This could be done for instance using the hypervolume indicator.
- Further research is needed to compare MOPRISM with other surrogate approaches. Comparisons should be made with the approach proposed by Liu, et al (2008) and the approach proposed by Fu, *et al.* (2008) named ParEGO. The approaches should be tested using numerical test functions and high dimensional objective optimization problems, and indicators for the reduction of computing time and convergence criteria should be evaluated.
- Further research could include a combination of the two general approaches that is; an algorithm that combines the benefits of the surrogate model optimization and which at the same time is parallelized and runs in a parallel computing infrastructure like a virtual cluster in the Cloud.

The Model-based Design and Control (MoDeCo) approach has great potential to find sustainable solutions for the management of urban wastewater. We hope that this research contribute to change the way that engineers design urban wastewater systems and change the way of thinking of urban water managers towards a holistic approach.

8 Appendix

8.1 History of urban wastewater systems

The summary of the evolution of UWwS is presented below together with a chronological time line scheme (Figure 8.1).

- **Early history times: 4000BC – 450 AC**

Three common factors can be identified in the early historical times: i) the need for proper disposal of human wastes was not fully understood, but there was recognition of some of the benefits (fewer odours for example) of taking these wastes away from the home; ii) the sewer systems were built to convey rainwater but ended up in combined sewerage systems; and iii) even though the sewers were successful in their function they were constructed in a trial and error process and therefore it was not the ideal sewer-design strategy. Sewers have been known for centuries, as early as 4500 BC in Babylon clay was moulded into pipes and used to convey run-off wastewater. In ancient Roman Empire, settled communities used open canals to transport excess of rainfall run-off out of the settlements in order to avoid nuisance and damage to properties. Later on, those canals were also used to convey wastewater from the public baths and latrines and were covert to avoid bad odours. Thus, eventually was created the first sewer system called "Cloaca Maxima" (510 BC) that was used to convey wastewater from the city of Rome to the Tiber River (Burian, *et al.* 1999).

- **The sanitary dark ages: 450 - 1750**

For the most part, the construction of sewer systems up to the 1700s lacked proper engineering design and was conducted in piecemeal fashion. In addition to the inadequate design and construction practices, maintenance and proper operation of the systems were virtually neglected. In summary, many of the sewer systems of European urban areas during the 1600s and 1700s were grossly under-planned, poorly constructed, and inadequately maintained by the today's standards, resulting in poorly functioning systems with repeated blockages and frequent nuisance conditions (Burian, *et al.* 1999).

- **The age of sanitary enlightenment: 1750 – 1800s**

Although research uncovered the connection between polluted waters and disease, wastewater treatment was not widely practiced. The debate centred on whether it was more economical to treat the wastewater prior to discharge or treat the water source before distribution as potable water. Most of the cities used the dilution theory ignoring impacts to the recreational and the habitat of the receiving water. Formula methods were generally used (Roe, McMath, Adams), in spite of the fact that the results given by them lack consistency and are very erratic and unreliable. Even though the rational method was already introduced, the lag time in technology transfer limited its use and the design and construction of sewerage systems ended up in a poor functioning of the sewer systems.

- **The age of environmental awareness and technical development: 1900s – 2000s**

The development of the discipline has provided new knowledge, and new tools available for sustainable urban water management. Computational aids, like integrated models advanced during the last decade, have improved the planning, design, operation and control of urban drainage systems significantly. However, the integrated approach has not found wider application in practice. Some of the reasons are the slow transfer of new techniques to the practitioners, the highly uncertainty in the predictions and the lack of knowledge that still exists in some of the components e.g. the ecological effects of urban drainage caused by chemicals used in the society. It is also noticed that real time control has not found wider application (Harremoes 2002).

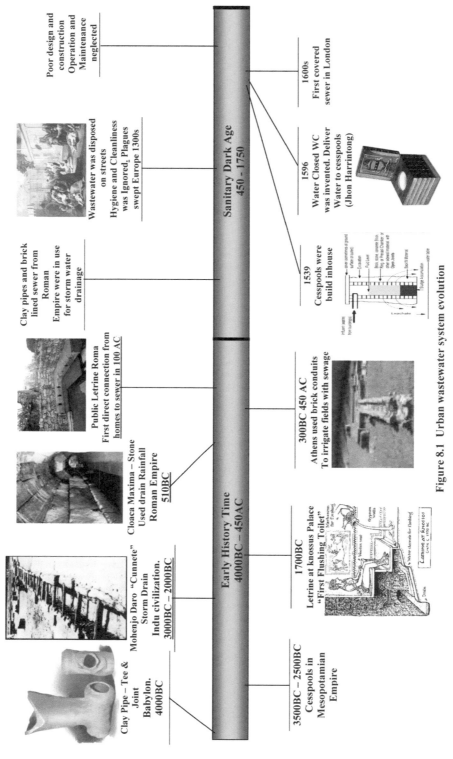

Figure 8.1 Urban wastewater system evolution

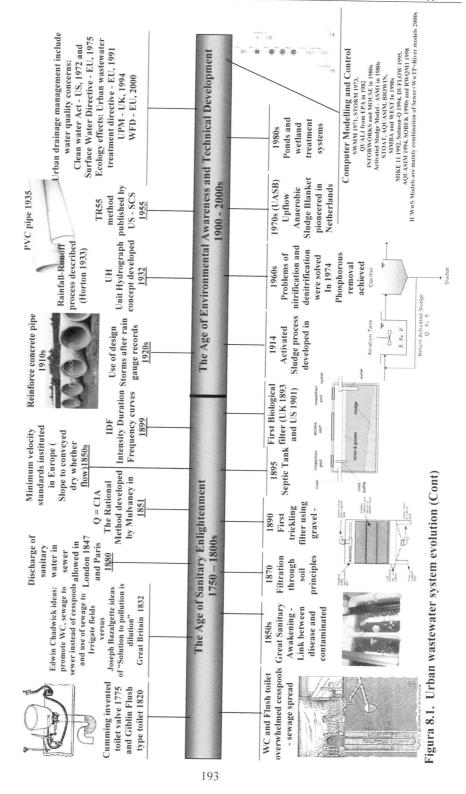

Figura 8.1. Urban wastewater system evolution (Cont)

Table 8.1. Considerations of urban wastewater system models

Modelling Water Quantity	Modelling Water Quality
Sewer Network	**Sewer Network**
Rainfall – Runoff quantity models reached high detail levels and nowadays the most diffuse commercial software can count on a wide range of possible approaches that can be sized to the characteristics of the catchment and to the availability of data	Pollution dynamic on the catchment surface is usually analysed by the use of conceptual approaches and empiric formulations; the fundamental reason is connected with the complexity of the pollution propagation/transformation processes and with the lack of experimental data
Rainfall – Runoff quantity models are mostly conceptual-distributed and their theoretical basis is characterised by linear/non-linear reservoir approach and channel approach.	Nevertheless all the most diffuse models demonstrated good adaptability to different catchment conditions and relatively easy calibration procedures giving reliable and robust results
Hydrologic depletions are based on the most diffuse either conceptual or physical approaches; commercial software generally allows the user for the selection of the most appropriate model for the specific case	Pollution/sediment transport analysis in sewers is still characterised by a low level of physical detail; only a few models simulate deposition/resuspension processes in the drainage system that is usually necessary for evaluating reduction in pipe hydraulic capacity
Flow propagation in the drainage system is usually performed by the use of 1D Saint Venant Equations either in the cinematic simplification of the full dynamic approach; most diffuse commercial models are characterised by quite old mathematical solvers even if optimised for using high performance processors.	A wide range of structures can be simulated (pump stations, storage, weirs and overflows, orifices, RTC, etc.) but very few models simulates detailed quality aspects of flow controls (sediment deposition; pollution abatement; etc.)
Full dynamic 1D S. Venant are usually used only for storm event simulation while cinematic approach is suitable for long term simulation of historical rainfall series	**What is still missing:** More details in build-up/wash-off quality analysis
A wide range of structures can be simulated (pump stations, storage, weirs and overflows, orifices, RTC, etc.) even if only few models are able to interact with them during simulations (manually start/stop pump stations; regulate overflows, etc.); for this reason models are quite useful for system design or verification but they are sufficiently adequate for supporting system management	More detailed approach for pollution propagation in sewers
What is still missing: Effective rainfall models are still too simple for managing complex hydrological structures such as snow-melting and overland flow on highly disturbed soils	Only a few models are adequate for managing highly distributed information coming from GIS databases (Digital Elevation Models; cellular/distributed hydrological parameters, etc.)
Only a few models are adequate for managing highly distributed information coming from GIS databases (Digital Elevation Models; cellular/distributed hydrological parameters, etc.)	The possibility for quality simulation of flow control structures
Available data are often insufficient for distributed models calibration - Approaches for pipe surcharge and surface flooding analysis are generally too simple and need to be improved	Available data are often insufficient for distributed models calibration
Interaction during simulation is usually not possible or limited to only a few of operational parameters	Interaction during simulation is usually not possible or limited to only a few of operational parameters

Based on the review of (Freni, *et al.* 2003), (Rauch, *et al.* 2002), (Harremoës and Rauch 1999), (Harremoes 2002), and (Shanahan, *et al.* 1998).

Table 8.1 Considerations of urban wastewater system models. (Cont)

Modelling Water Quantity	Modelling Water Quality
WwTP	WwTP
The modelling of the wastewater treatment subsystem is quite different from the modelling of sewer or river systems in two respects: first, the underlying hydraulics can nearly always be approximated crudely and, second, the modelling is built up around unit processes.	The commonly accepted model structure (Activated Sludge Model No1 – ASM1 – Petersen Matrix) is the basis for most available computer programmes. The organic mater is represented in terms of Chemical Oxygen Demand (COD).
Only in very particular cases flow propagation through reactors is modelled explicitly and even then in a simplified way, using transfer functions or the variable volume tank approach. Usually, instantaneous flow propagation is assumed (the outflow rate is assumed to be equal to the inflow rate at any time).	Mixing is typically modelled using the continuously stirred tank reactors (CSTR) in series approach. This approach allows to reasonably mimic the advection and dispersion of matter in different unit processes
Simulation of WwTP has found wide application in analysis of performance, bad less so in design; where old design method still prevail.	The main components modelled are based in the conventional technologies: activated sludge tanks, settlers, trickling filters, biofilms and anaerobic digestions
What is Still missing Integration with urban drainage and receiving water systems models that, at the moment, can only exchange input/output time series.	**What is Still missing** The most difficult issue is related to the function of the clarifier and relation to the settling properties of the sludge. This cannot be modelled adequately for design and the model has to be adaptable to data on the settlability in real time control
Source codes are generally not available and for this reason customization is possible only within the limits of the software (no input/output format change is possible; user defined variables can not be calculated)	Low development or lack of knowledge in the modelling of no-conventional treatment process like ponds and wetlands.
The number of functions and parameters is so large that it is impossible to calibrate in each particular case. The question is whether treatment plants are so uniform in phenomena, function that most functions and parameters are universal, and do not need calibration.	The fractionation of the organic mater in terms of COD represent a limitation to couple with Sewer and River models that are based in Biological Oxygen Demand (BOD).
Rivers	**Rivers**
Quantity models are generally similar to the drainage models. Flow of water in a river is described by the continuity and momentum equations. The latter is known as the Navier-Stokes or Reynolds equation. Complex models are available but for water quality purposes mostly the well-known, cross-sectionally integrated (1D) Saint Venant equations or approximations to these equations are use.	Quality models generally adopt conceptual approaches and have quite fast computational routines in order to perform long term simulations
Also complex structures such as bridges, culverts, reservoirs, dam breaks, flood plain, etc. could be modelled.	Models include the water quality changes in rivers due to physical transport and exchange processes (advection and diffusion/dispersion equation) and biological, biochemical or physical conversion processes (Conversion equations).
What is Still missing To create models which are compatible with the existing standard models for describing wastewater treatment and can be straightforwardly linked to them.	**What is Still missing** It is difficult to simulate impacts in receiving systems with a reasonable degree of fit, even for the conventional parameters (OD and Ammonia).
The gap between quantity and quality models is generated by the difficulties in understanding quality processes and in the lack of data for model approach verification	A very fundamental concern with the existing approach is the fact that using BOD as a state variable and poor representation of benthic flux terms. Intrinsically means that mass balances cannot be closed.
	Prediction of ecological effects of urban drainage is not possible by modelling, because of lack of knowledge related to cause effect relationships.

Based on the review of (Freni, et al. 2003), (Rauch, et al. 2002), (Harremoës and Rauch 1999), (Harremoes 2002) and (Shanahan, et al. 1998)

8.2 Case Study of Cali

Table 8.2 Sub-catchment descriptions of Sector 1A, expansion zone of Cali

District/ Catchment	Area	width	slope	Discharge Reach	Population Density	Water consumption	Water Losses	Average		Wastewater production peak factor	Maximum	Population	DWF
	ha	m	%	Nodo - Nodo	h/ha	l/h.d	l/h.d	l/h.d	l/s.ha		l/s.ha	h	l/s
1	1.4	101	1.144	1 - 2	310	220	44	176	0.63	1.7	1.07	434	1.5
2	0.78	47	1.218	1 - 2	310	220	44	176	0.63	1.7	1.07	242	0.8
3	1.9	106	0.89	2 - 3	310	220	44	176	0.63	1.7	1.07	589	2.0
4	2.3	104	0.887	3 - 4	310	220	44	176	0.63	1.7	1.07	713	2.5
5	2.6	96	0.9	3 - 4	310	220	44	176	0.63	1.7	1.07	806	2.8
6	0.98	69	0.93	4 - 5	310	220	44	176	0.63	1.7	1.07	304	1.1
7	0.9	62	1	4 - 5	310	220	44	176	0.63	1.7	1.07	279	1.0
8	1.29	85	1.026	5 - 6	310	220	44	176	0.63	1.7	1.07	400	1.4
9	1.23	72	1.3	16 - 17	310	220	44	176	0.63	1.7	1.07	381	1.3
10	2.03	154	1.34	16 - 17	310	220	44	176	0.63	1.7	1.07	629	2.2
11	1.05	74	0.6	7 - 8	310	220	44	176	0.63	1.7	1.07	326	1.1
12	0.83	65	1.14	7 - 8	310	220	44	176	0.63	1.7	1.07	257	0.9
13	1.67	126	1.7	8 - 9	310	220	44	176	0.63	1.7	1.07	518	1.8
14	1.44	97	0.99	9 - 10	310	220	44	176	0.63	1.7	1.07	446	1.5
15	2.29	90	1.047	11 - 12	310	220	44	176	0.63	1.7	1.07	710	2.5
16	2.39	85	1.047	12 - 13	310	220	44	176	0.63	1.7	1.07	741	2.6
17	1.39	41	0.05	20 - 13	310	220	44	176	0.63	1.7	1.07	431	1.5
18	0.9	89	1.8	7 - 8	310	220	44	176	0.63	1.7	1.07	279	1.0
19	2.45	131	0.9	8 - 9	310	220	44	176	0.63	1.7	1.07	760	2.6
20	3.59	148	1.22	9 - 10	310	220	44	176	0.63	1.7	1.07	1113	3.9
21	2.85	162	0.5	11 - 12	310	220	44	176	0.63	1.7	1.07	884	3.1
22	1.5	190	0.65	13 - 21	310	220	44	176	0.63	1.7	1.07	465	1.6
23	4.66	254	1.3	21 - 22	310	220	44	176	0.63	1.7	1.07	1445	5.0
24	2.42	241	0.6	21 - 22	310	220	44	176	0.63	1.7	1.07	750	2.6
25	4.05	170	0.7	14 - 15	310	220	44	176	0.63	1.7	1.07	1256	4.3
26	2.7	212	0.61	16 - 17	310	220	44	176	0.63	1.7	1.07	837	2.9
27	2.7	132	0.6	14 - 15	310	220	44	176	0.63	1.7	1.07	837	2.9
28	2.37	122	0.618	15 - 16	310	220	44	176	0.63	1.7	1.07	735	2.5
29	2.08	104	0.627	16 -17	310	220	44	176	0.63	1.7	1.07	645	2.2
30	3.95	186	0.61	23 - 24	310	220	44	176	0.63	1.7	1.07	1225	4.2
31	4.65	246	0.6	23 - 24	310	220	44	176	0.63	1.7	1.07	1442	5.0
32	2.64	176	0.57	25 - 26	310	220	44	176	0.63	1.7	1.07	818	2.8
Total	69.98											21694	75.1

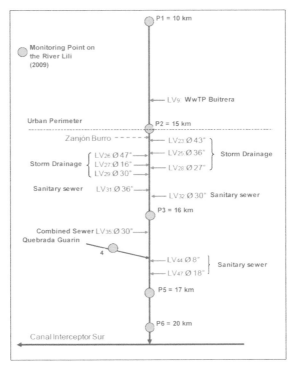

Figure 8.2 Discharges to the River Lili

Source: (DAGMA and UNIVALLE 2009)

Table 8.3 Combined sewer design for sector 1A – Cali.

Section		From area, branch & heavy consumers	Dry weather flow								Wet Weather Flow							
			Tributary area		Peak domestic waste water			Infiltration		Peak Sanitary Flow	Tributary area		Runoff coefficient			Estimated time of concentration	Average rainfall intensity (from IDF curve)	Storm Flow
From	To		Increment	Cummulative	Rate	Increment	Cummulative	Increment	Total	Q_s	Increment	Cummulative	$\psi^{(1)}$	$A_r = \psi^*A$	Cummulative	t_c	i'_{t_c}	$Q_l = i'_{t_c} \Sigma A_r$
			A	ΣA	q_s	q_s^*A	Σq_s^*A				A	ΣA	C		ΣA_r			
No	No		ha	ha	L/s*ha	L/s	L/s	L/s	L/s	L/s	ha	ha		ha	ha	min	L/s*ha	L/s
1	2	3	4	5	6	7	8	9	10	11	12	13	14	15	16	17	18	19
N1	N2	1, 2	2.18		1.07	2.34				2.3	2.18		0.50	1.09		<10	323	352
N2	N3	3	1.90	4.08	1.07	2.04	4.38			4.4	1.90	4.08	0.50	0.95	2.04	<10	323	658
N3	N4	4, 5	4.9	8.98	1.07	5.26	9.64			9.6	4.90	8.98	0.50	2.45	4.49	<10	323	1448
N4	N5	6, 7	1.88	10.86	1.07	2.02	11.66			11.7	1.88	10.86	0.50	0.94	5.43	<10	323	1751
N5	N15	8	1.29	12.15	1.07	1.38	13.04			13.0	1.29	12.15	0.50	0.65	6.08	<10	323	1960
N6	N7	11, 12, 18	2.78		1.07	2.98				3.0	2.78	0.00	0.50	1.39		<10	323	448
N7	N8	13, 19	4.12	6.90	1.07	4.42	7.41			7.4	4.12	6.90	0.50	2.06	3.45	<10	323	1113
N8	N9	14, 20	5.03	11.93	1.07	5.40	12.81			12.8	5.03	11.93	0.50	2.52	5.97	<10	323	1924
N9	N10	-	0	11.93	1.07	0.00	12.81			12.8	0.00	11.93	0.50	0.00	5.97	<10	323	1924
N10	N11	15, 21	5.14	17.07	1.07	5.52	18.32			18.3	5.14	17.07	0.50	2.57	8.54	<10	323	2753
N11	N19	16	2.39	19.46	1.07	2.57	20.89			20.9	2.39	19.46	0.50	1.20	9.73	<10	323	3138
N12	N13	25, 27	6.75		1.07	7.25				7.2	6.75	0.00	0.50	3.38		<10	323	1089
N13	N14	28	2.37	9.12	1.07	2.54	9.79			9.8	2.37	9.12	0.50	1.19	4.56	<10	323	1471
N14	N22	26, 29	4.78	13.90	1.07	5.13	14.92			14.9	4.78	13.90	0.50	2.39	6.95	<10	323	2242
N15	N16	-	12.15		1.07	13.04				13.0	12.15	0.00	0.50	6.08		10	323	1960
N16	N17	9, 10	3.26	15.41	1.07	3.50	16.54			16.5	3.26	15.41	0.50	1.63	7.71	11	312	2403
N17	N18	-	0	15.41	1.07	0.00	16.54			16.5	0.00	15.41	0.50	0.00	7.71	13	293	2256
N18	N19	17	1.39	16.80	1.07	1.49	18.04			18.0	1.39	16.80	0.50	0.70	8.40	14	284	2388
N19	N20	22	20.96	37.76	1.07	22.50	40.54			40.5	20.96	37.76	0.50	10.48	18.88	15	276	5215
N20	N21	23, 24	7.08	44.84	1.07	7.60	48.14			48.1	7.08	44.84	0.50	3.54	22.42	16	269	6023
N21	N22	-	0	44.84	1.07	0.00	48.14			48.1	0.00	44.84	0.50	0.00	22.42	17	262	5864
N22	N23	-	13.90	58.74	1.07	14.92	63.06			63.1	13.90	58.74	0.50	6.95	29.37	17	262	7682
N23	N24	30, 31	8.6	67.34	1.07	9.23	72.29			72.3	8.60	67.34	0.50	4.30	33.67	18	255	8581
N24	N25	-	0	67.34	1.07	0.00	72.29			72.3	0.00	67.34	0.50	0.00	33.67	19	249	8368
N25	N26	32	2.64	69.98	1.07	2.83	75.12			75.125	2.64	69.98	0.50	1.32	34.99	20	254	8893

Conduit Design

Total Flow $Q = Q_s + Q_r$	Elevation				Length	Slope of sewer	Profile & size	Capacity $(Q = V_o^*A)$	Velocity (Colebrook White)	Wet Wether Peak flow		Dry Wether Peak flow		Flow time		
	Ground surface		Invert							Q_t/Q_o	Velocity $= v_o * v/v_o$ (v/v_o from Table)	Q_s/Q_o	Velocity $= v_o * v/v_o$ (v/v_o from Table)	Increment	Total	
	Upper end	Lower end	Upper end	Lower end	L			Q_o	v_o					$t = L/v$	Inlet time is 5 min	
L/s	m	m	m	m	m	o/oo	m	L/s	m/s	%	m/s	%	m/s	s	s	min
20	21	22	23	24	25	26	27	28	29	30	31	32	33	34	35	36
354	980.00	976.24	978.55	974.79	160	23.51	0.45	441	2.77	0.80	3.02	0.005	0.83	53	353	5.88
662	976.24	974.44	974.56	972.77	185	9.71	0.68	826	2.31	0.80	2.51	0.005	0.69	74	427	7.11
1458	974.44	971.531	972.62	969.71	193	15.08	0.83	1747	3.27	0.83	3.56	0.006	0.98	54	481	8.01
1763	971.531	969.94	969.53	967.94	151	10.52	1.00	2419	3.08	0.73	3.30	0.005	0.92	46	527	8.78
1973	969.94	968.226	967.94	966.23	117	14.68	1.00	2858	3.64	0.69	3.85	0.005	1.09	30	557	9.28
451	975	973.26	973.40	971.66	150	11.63	0.60	663	2.35	0.68	2.47	0.005	0.70	61	361	6.01
1120	973.26	970	971.58	968.33	138	23.59	0.68	1288	3.60	0.87	3.91	0.006	1.08	35	396	6.60
1937	970	967.91	968.10	966.01	136	15.40	0.90	2219	3.49	0.87	3.78	0.006	1.05	36	432	7.20
1937	967.91	967.10	965.91	965.10	75	10.75	1.00	2445	3.11	0.79	3.38	0.005	0.93	22	454	7.57
2771	967.10	965.00	965.00	962.90	200	10.50	1.10	3105	3.27	0.89	3.52	0.006	0.98	57	511	8.51
3159	965.00	964.40	962.65	962.05	112	5.35	1.35	3794	2.65	0.83	2.89	0.006	0.80	39	550	9.16
1096	965.00	964.45	963.00	962.45	136	4.06	1.00	1503	1.91	0.73	2.05	0.005	0.57	66	366	6.11
1481	964.45	963.26	962.45	961.26	151	7.86	1.00	2092	2.66	0.71	2.83	0.005	0.80	53	420	6.99
2257	963.26	962.07	961.16	959.97	143	8.36	1.10	2771	2.92	0.81	3.18	0.005	0.87	45	465	7.74
1973	968.226	966.57	966.23	964.57	174	9.49	1.00	2299	2.93	0.86	3.18	0.006	0.88	55	612	10.20
2420	966.57	964.94	964.47	962.84	187	8.76	1.10	2836	2.98	0.85	3.25	0.006	0.90	58	669	11.16
2273	964.94	964.90	962.54	962.15	176	2.22	1.40	2690	1.75	0.84	1.90	0.006	0.52	92	762	12.70
2406	964.90	964.40	962.15	961.75	161	2.45	1.40	2823	1.83	0.85	2.00	0.006	0.55	81	842	14.04
5255	964.40	963.47	961.75	960.97	116	6.73	1.50	5611	3.18	0.94	3.34	0.007	0.95	35	877	14.62
6071	963.47	962.39	960.87	959.79	136	7.93	1.60	7213	3.59	0.84	3.91	0.007	1.08	35	912	15.20
5912	962.39	962.07	959.39	959.07	150	2.17	2.00	6774	2.16	0.87	2.34	0.007	0.65	64	976	16.27
7745	962.07	960.90	959.07	957.90	167	6.97	2.00	12129	3.86	0.64	4.01	0.005	1.16	42	1018	16.96
8653	960.90	960.00	957.90	957.00	176	5.12	2.00	10395	3.31	0.83	3.60	0.007	0.99	49	1067	17.78
8440	960.00	959.52	956.75	956.27	175	2.72	2.25	10313	2.59	0.82	2.83	0.007	0.78	62	1129	18.81
8968	959.52	959.20	956.12	955.80	156	2.06	2.40	10620	2.35	0.84	2.56	0.007	0.70	61	1190	19.83

Table 8.4 Design of CSO setting and on-line storage

Overflow Setting using different methods

__Basic data__

DWF	0.075 m3/s	
	75.1 l/s	6490784.96 l/d
Area	70 ha	
Population	21694 hab	

a) Setting = 6*DWF

CSO Setting	451 l/s	CSO Setting

b) Formula A

Setting = DWF + 1360P + 2E   l/d
DWF = PG + I + E
P: population
G: water consumption per person (l/d)
I: pipe infiltration rate
E: average industrial effluent (l/d)
Asumption E=0

CSO Setting	417 l/s	Equivalent to	5.5 *DWF

c) River water quality adjacent to the overflow limited to

COD = 20 mg/l for the 2 year return period, 20 minute duration event.

COD Limit	20 mg/l	Asumption
CSO COD	856 mg/l	
Lili River Q	386 l/s	Flow exceeded 50% of the time Lili Duration Curve
Lili River COD	11 mg/l	
WWF	8893 l/s	
Qoverflow	4.2 l/s	Maximum allowed to not pass the COD Limit
	7803 =	7803

CSO setting	457 l/s	Equivalent to	6.1 *DWF

Volume of Storage Tank using different methods

a. Scottish Development Department (SDD) Recommendation

Dilution	Storage capcity (l)	
6 to 1	40 P	
4 to 1	40 P	
2 to 1	80 P	
1 to 1	120 P	
QminRiver	140 l/s	flow rate exceeded 95% of time
DWF	75.1 l/s	
__Qriver/DWF__	2 :1	
Storage Vol =	80 P	from table above
__Storage Vol__	1736 m3	
Storage Vol =	120 P	
__Storage Vol__	2603 m3	maximum volume proposed

Ojo the flow in the river is 386l/s less
than the 8.9 m3/s of CSO so dilution is
lower than 1 for critical flow in the river

This CSO desing is in combination with Formula A. So:

the setting flow is	417 l/s	Equiv	5.5 xDWF
Storage Vol =	120 P		
and include a storage volume of	2603 m3		
Max depth	2.5 m		
__Functional Area (Max h=2.5)__	1041 m2		

Table 8.5 Design of the wastewater treatment plant for Sector 1A – Cali.

Design Of Wwtp Using Meltcaf And Eddy Process (For One Lane) Modified Ludzack and Ettinger (MLE) configuration (BOD and N removal) (Include Anoxic + Aerated + Settler)

ASM1		Parameter	Value	Unit	Formula
		BIOREACTOR VOLUME (AERATION TANK)			
Notation	**SRT**	Solid retention time	12.7	days	
	Q	Flow rate	6.5	ML/d	
YH	**Y**	Biomass yield for Heterotrophos (VSS and S ratio)	0.4	kgVSS/Kg S	
	BOD	Biochemical oxygen demand concentration	228	mg/L	
	BOD	BOD Load	1,479	kg/d	L_{BOD} = BOD concentration*flowrate
Ss+Xs	**So**	Influent substrate - Biodegradable COD in the influent (bCOD)	365	mg/L	So =1.6*BOD
	S	Actual or effluent substrate (BOD or sbCOD)	5	mg/L	
	T	Minimum temperature for design	18	C	
bH$_{20}$	**K$_{d20}$**	Endogenous decay coefficient for heterotrophos at 20°C	0.12	gVSS/ g VSS/d	
bH	**K$_d$**	Endogenous decay coefficient for heterotrophos at the design temperature	0.11095	d^{-1}	$Kd = Kd_{20C} * \Theta^{T-20}$
	Θ	Temperature coefficient for kd and kdn	1.04	-	
XBH	**A**	Heterotrophic biomass production	388	kg/d	A = Q*Y*(So-S)/(1+k$_d$*SRT)
fP	**fd**	Fraction of cell mass remaining as cell debris (yielding to particulate products)	0.15	g/g	
Xp	**B**	Cell debris (particulate produces from biomass decay)	82	kg/d	B = fd*kd*SRT*A
YA	**Yn**	Biomass yield for autotrophs (nitrifiers)	0.12	gVSS/gNH$_4$-N	
SNH	**NH$_x$**	Concentration of NH4-N in the influent flow that is nitrified	46.9	mgN/L	NO$_x$ = 0.8* TKNo
	TKNo	Total Kjeldahl Nitrogen concentration	59	mgN/L	
	TKNo Load	Total Kjeldahl Nitrogen load	380	kgN/d	L_{TKN} = TKN concentration*flowrate
bA$_{20}$	**Kdn$_{20C}$**	Endogenous decay coeffiecient for autotrophs at 20 C	0.08	gVSS/gVSS/d	
bA	**Kdn**	Endogenous decay coeffiecient for autotrophs	0.074	d^{-1}	$Kdn = Kdn_{20C} * \Theta^{T-20}$
XBA	**C**	Autotrophic (Nitrifying bacteria) biomass production	19	kg/d	
	TSS load	Total suspended solids load	1272	Kg/d	
	TSS	Total suspended solids concentration	196	mg/L	
	VSS	Volatile suspended solids concentration	167	mg/L	VSS = 0.85*TSS
	nbVSS	Non-biodegradable Volatile Suspended Solids in the influent	55	mg/L	nbVSS = 0.33*VSS
Si	**D**	Non-biodegradable VSS in influent	357	kg/d	D = Q * nbVSS
Xi	**E**	Inert TSS in influent	191	kg/d	E = Q*(TSSo-VSSo)
	P$_{X,TSS}$	TSS production	1122	kg/d	P$_{X,TSS}$ = (A/0.85) + (B/0.85) + (C/0.85) + D + E
	X$_{TSS}$	TSS in the bioreactor (= MLSS)	4,000	mg/L	
	V	**Volume of the reactor**	3.56	ML	V = P$_{X,TSS}$ * SRT / X$_{TSS}$
		Adopting depth = 4m ==> Area =	890.6	m2	

Table 8.5 Design of the wastewater treatment plant for Sector 1A – Cali (Cont.)
Design Of Wwtp Using Meltcaf And Eddy Process (For One Lane)
Modified Ludzack and Ettinger (MLE) configuration (BOD and N removal)
(Include Anoxic + Aerated + Settler)

ASM1		Parameter	Value	Unit	Formula
Notation	**ANOXIC VOLUME (N-REMOVAL)**				
	Q	Flowrate	6.5	ML/d	
	A	Heterotrophic biomass	388	kg/d	
	B	Cell debris	82	kg/d	
	C	Autotrophic (Nitrifying bacteria) biomass	19	kg/d	
	$P_{X,bio}$	Biomass as VSS wasted (parts A,B and C of $P_{X,VSS}$ equation)	488	Kg/d	$P_{X,bio} = A + B + C$
	$N_{e,ammo}$	Effluent NH_4-N concentration	0.5	mg/L	
	TKNo	Influent TKN concentration	59	mgN/L.	
SNO	NO_x	Nitrate produce in aerated zone, expressed as a concentrationrelative to inflow	49	mg/L	$NOx = TKNo-N_{e,ammon}-(0.12*P_{X,bio}/Q)$
	TN	Total nitrogen effluent	12	mg/L	$TN = N_{e,ammon} + Organic-N + N_e$
	Organic-N	Unbiodegradable organic nitrogen in effluent	2.5	mg/L	
	N_e	Nitrate effluent	9.0	mg/L	$N_e = TN - N_{e,ammon} - Organic-N$
	R	RAS recycle ratio (RAS flowrate/influent flowrate)	0.6	unitless	
	IR	Internal recycle ratio (MLR flowrate/ influent flowrate)	3.85	unitless	$IR = (NO_x/N_e) - 1 - R$
	NO_3-N	Amount of NO_3-N fed to the anoxic tank	260	kg/d	$NO_3-N = N_e * Q * (IR+R)$
	V_{nox}	**Volume of anoxic tank**	**1.00**	ML	
		Adopting depth = 4m ==> Area =	250.0	m2	
	SRT aer	Sludge retention time in aeration tank	12.7	days	
	V_{aer}	Volume aeration tank	3.6	ML	
	Kd	Endogenous decay coefficient for the design temperature	0.11095	d^{-1}	
	Y	VSS and S ratio	0.400	kgVSS/Kg S	
	So	Biodegradable COD on the influent	365	mg/L	$So = bCOD = 1.6*BOD$
	S	Typical target effluent for BOD concentration	5	mg/L	
	Xb	Active mass concentration in aeration tank (same in anoxic zone)	1,382	mg/L	$Xb = Q * SRT_{aer} * Y * (So - S)/(V_{aer}*(1+Kd*SRT_{aer}))$
	BODo	BOD concentration in the influent	228	mg/L	
	F/M_b	BOD F/M ratio based on active biomass concentration	1.07		$F/M_b = Q*BODo/(V_{nox}*X_b)$
	COD	Chemical oxygen demand	333.90	mg/L	
Ss	rbCOD	Readily bio degradable COD	83.48	mg/L	$rbCOD = 0.25 * COD$
Ss+Xs	bCOD	Bio degradable COD	364.80	mg/L	$bCOD = So = 1.6* BOD$
	rbCOD/bCOD		0.23	unitless	rbCOD/bCOD
	$SDNR_{20C}$	Specific denitrification rate at 20 C	0.20	g NO_3-N/gbiomass*d	
	Θ	Temperature coefficient for SDNR (specific denitrification rate)	1.026	unitless	
	SDNR	Specific denitrification rate	0.19	g NO_3-N/gbiomass*d	$SDNR_T = SDNR_{20C} * \Theta^{T-20}$
	NO_x	Nitrate removed	263	kg/d	$NO_x = V_{nox} * SDNR*X_b$
	Anoxic V fraction		21.92	%	Anoxic V fraction = $V_{nox}/(V_{aer} + V_{nox}) * 100$

ALKALINITY

	Alk_0	Influent alkalinity	200	mg $CaCO_3$/L	
	$Alk_{used,nitr}$	Alkalinity used in nitrification	7.14	g/gN nitrified)	
	$Alk_{used,nitr}$	Alkalinity produced in nitrification	350.09	kg/d	$Alk_{used,nitr} (Kg/d)= NO_x * Alk_{used,nitr} (Kg/d)$
	$Alk_{used,nitr}$		53.95	mg/L	$Alk_{used,nitr} (mg/L)= Alk_{used,nitr} (Kg/d) / Q (ML/d)$
	$Alk_{prod,denitr}$	Alkalinity produced in denitrification	3.57	g/gN denitrified)	
	$Alk_{prod,denitr}$		937.30	kg/d	$Alk_{prod,denitr} (Kg/d)= (NO_x-N_e) * Alk_{prod,denitr} (Kg/d)$
	$Alk_{prod,denitr}$		144.45	mg/L	$Alk_{prod,denitr} (mg/L)=Alk_{prod,denitr} (Kg/d) / Q (ML/d)$
	Alk_{eff}	Alkalinity effluent	290	mg/L	$Alk_{eff} = Alk_0 - Alk_{use,nitr} + Alk_{prod,denitr}$
	Alk_{added}	Alkalinity added to reach 80 mg/L.	NO NEED TO ADD ALKALINITY	Kg/d	$Alk_{added} = (80 - Alk eff) * Q$

Table 8.5 Design of the wastewater treatment plant for Sector 1A – Cali (Cont).
Design Of Wwtp Using Meltcaf And Eddy Process (For One Lane)
Modified Ludzack and Ettinger (MLE) configuration (BOD and N removal)
(Include Anoxic + Aerated + Settler)

AERATION CAPACITY				
Parameter		Value	Unit	Formula
Q_{in}		6.5	ML/d	
$C_{BOD,IN}$		228	mg/L	
$C_{BOD,OUT}$		5	mg/L	
L_{BOD}		1,447	kg/d	$L_{BOD} = Q*(C_{BOD,IN} - C_{BOD,OUT})$
a	Factor from nitrogen removal theory	0.5		
COR		723.48	kg/d	$COR = a * L_{BOD}$
c	Factor from nitrogen removal theory	4.57		
$C_{NH4,in}$		46.85	mg/L	$C_{NH4,in} = 0.8 * TKN$
$C_{NH4,out}$		0.5	mg/L	
$L_{ammonia\ nitrified}$		300.8	kg/d	$L_{ammonia\ nitrified} = Q*(C_{NH4,in} - C_{NH4,out})$
NOR		1,374.45	kg/d	$NOR = c * L_{ammonia\ nitrified}$
d	Factor from nitrogen removal theory	2.86		
DOR	O_2 released from denitrification	576	kg/d	$DOR = d * (NO_x - N_e * Q)$
b	Factor from nitrogen removal theory	0.1	d^{-1}	
X	MLSS (biomass concentration in bioreactor)	4,000	mg/L	
EOR	Endogenous oxygen requirement	1,425	kg/d	$EOR = b * X * V$
AOR	Actual oxygen requirement	2,947	kg/d	$AOR = COR + EOR + NOR - DOR$
AOR	Actual oxygen requirement	122.80	kg/h	
$T_{aeration}$	Temperature for calculating aeration	22	C	
C_{ST}	Table value for dissolved oxygen at temperatute T, at surface level	8.73	mgO_2/L	
h	Submersion depth of the diffusers	4.25	m	
$C*_\infty$	Steady state dissolved oxygen saturation concentration attained at infinite time at water temperature T and field atmosferic pressure	9.95	mgO_2/L	$C*_\infty = C_{ST}*(1 + 0.035 * (h-0.25))$
α	Ratio of the oxygen mass transfer coefficients measured in sewage and in clean water	0.90		$\alpha = k_L a_{sewage}/k_L a_{clean\ water}$
β	Ratio of oxygen saturation concentrations in sewage and in clean water	0.98		$\beta = C*_{\infty,sewage}/C*_{\infty,clean\ water}$
C_{ST20}	Table value for dissolved oxygen at temperatute 20 C at surface level	9.08	mgO_2/L	
$C*_{\infty,20}$	Steady state dissolved oxygen saturation concentration attained at infinite time at water temperature 20 C and standard atmosferic pressure	10.43	mgO_2/L	$C*_{\infty,20} = C_{ST20}*(1 + 0.035 * h)$
C_L	Actual oxygen concentration (Set point)	0.9	mgO_2/L	
Θ	Temperature correction coefficient	1.024		
SOTR	Standard oxigen transfer rate	153.30	kgO_2/h	$SOTR = (1/\alpha)*(C*_{\infty,20} * \theta^{(20-T)}* AOR) / (\beta * C*_\infty - C_L)$
SOTR	Standard oxigen transfer rate	3,679.29	kgO_2/d	
SOTE	Standard oxygen transfer efficiency	2.5	%/m	
SOTE	Standard oxygen transfer efficiency	10.625	%	$SOTE (\%) = SOTE (\%/m)*h (m)$
Q_{air}	Air flowrate	5,153.07	m^3/h	$Q_{air} = SOTR/(0.28*SOTE)$
$Q_{air,diffuser}$	Air flowrate per diffuser	3	m^3/h/diffuser	
$N_{diffuser}$	Number of diffusers	1,718	diffuser	$N_{diffuser} = Q_{air} / Q_{air\ diffuser}$

FINAL SETTLING TANK (Standard loading rate method)				
Parameter		Value	Unit	Formula
Q	Peak flowrate (used for designing the FST)	0.27	ML/h	
$OR_{standard}$	Standard overflow rate	1.25	$m^3/m^2/h$	
A	Total area	216	m^2	$A = Q(m^3/h)/OR_{standard}(m^3/m^2/h)$
G	Solids loading rate	8.0	$kg/h/m^2$	$G = \dfrac{M_X}{A} = \dfrac{X(Q+Q_r)}{A}$
n	Number of final settling tanks/lane	1	tank	
d	Diameter	16.6	m	$d = \sqrt{((4*A_T/n)/\pi)}$
	Adopting depth=4m ==> Volume =	864	m3	

Table 8.6 Results of the steady state simulation for the state variables of the model

Variable [mg/l]	Anoxic R2	Aerated R5	Effluent	Underflow
Ss	1.57	0.70	0.88	0.88
Xi	1590.09	1589.82	6.34	4082.59
Xs	50.34	23.06	0.09	59.39
Xbh	1553.95	1554.92	6.21	4003.53
Xba	120.97	122.53	0.49	315.35
Xp	557.20	561.08	2.24	1442.59
So	0.00	0.53	0.67	0.67
Sno	0.92	9.05	11.44	11.43
Snh	7.84	0.24	0.31	0.31
Snd	0.65	0.61	0.77	0.77
Xnd	3.69	1.82	0.01	4.68
Salk	1.69	0.57	0.72	0.72
TSS	2907.18	2889.92	11.53	7433.47

Table 8.7 Results of the steady state simulation for secondary settler

Settler	TSS [mg/l]
layer 10	11.40
layer 9	16.86
layer 8	27.61
layer 7	63.50
layer 6	312.37
layer 5	276.37
layer 4	275.16
layer 3	275.07
layer 2	275.04
layer 1	7311.22

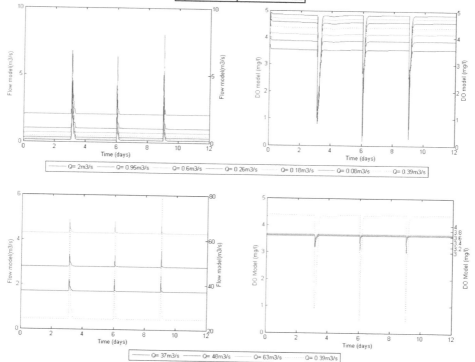

Figure 8.3 Sensitivity of dissolved oxygen modelled to different boundary flow conditions

Table 8.8 Pipe catalogue used for UWwS of Sector 1A - Cali

Index	Diameter (m)	Cost €/m (2010)
0	0.15	7.3
1	0.20	8.9
2	0.25	11.6
3	0.30	28.7
4	0.38	37.3
5	0.46	53.3
6	0.53	64.5
7	0.61	67.2
8	0.69	75.9
9	0.76	88.2
10	0.84	101.1
11	0.91	114.6
12	1	139.6
13	1.1	146.2
14	1.2	164.3
15	1.3	190.5
16	1.4	216.5
17	1.5	263.7
18	1.6	293.2
19	1.7	330.8
20	1.8	375.0
21	2	411.1
22	2.1	443.8
23	2.25	494.3
24	2.4	546.8
25	2.7	657.5
26	2.85	715.5
27	3	775.2
28	3.15	836.7
29	3.3	899.9
30	3.45	964.7
31	3.6	1031.1
32	4.05	1239.7

Note: The cost per pipe includes supply, transport and installation. No excavation and backfill are included.

Figure 8.4 Sensitivity of the crossover fraction of NSGA II algorithm for sewer network design

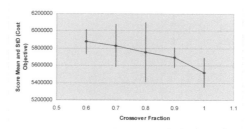

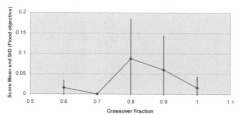
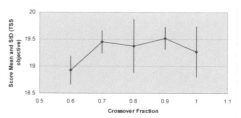

8.3 Case Study of Gouda

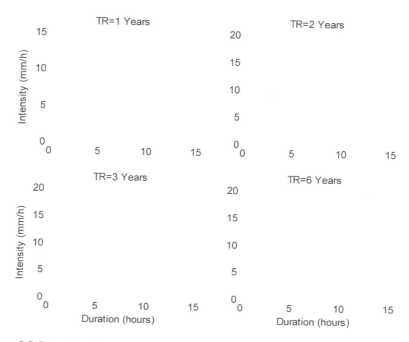

Figure 8.5 Intensity – Duration – Frequency (IDF) curves estimate for Gouda base on the 6 years of hourly precipitation data from radar.

Table 8.9 Indicators of goodness of fit for DWF modelled

Indicators of Goodness of Fit	Summer	Winter
mean squared error (mse)	8245	19003
normalised mean squared error (nmse)	0.5	0.8
root mean squared error (rmse)	90.8	137.9
normalised root mean squared error (nrmse)	0.7	0.9
mean absolute error (mae)	73.0	108.0
mean absolute relative error (mare)	0.1	0.1
coefficient of correlation (r)	0.8	0.8
coefficient of determination (r-squared)	0.7	0.6
coefficient of efficiency (e)	0.5	0.2
maximum absolute error	237	416
maximum absolute relative error	0.5	0.5

Table 8.10 Indicators of goodness of fit for WWF modelled

Indicators of Goodness of Fit	Low Freq	High Freq
mean squared error (mse)	221954	28525
normalised mean squared error (nmse)	0.4	0.8
root mean squared error (rmse)	471.1	168.9
normalised root mean squared error (nrmse)	0.6	0.9
mean absolute error (mae)	301.7	114.0
mean absolute relative error (mare)	0.2	0.2
coefficient of correlation (r)	0.8	0.8
coefficient of determination (r-squared)	0.7	0.6
coefficient of efficiency (e)	0.6	0.2
maximum absolute error	1483	748
maximum absolute relative error	0.5	1.4

Table 8.11 Correlation between variables for COD model tree

	P3	P2	P1	P	Q	COD
P3	1	0.35	0.14	0.06	0.34	-0.35
P2	0.35	1	0.23	0.16	0.49	-0.45
P1	0.14	0.23	1	0.19	0.69	-0.57
P	0.06	0.16	0.19	1	0.25	-0.28
Q	0.34	0.49	0.69	0.25	1	-0.92
COD	-0.35	-0.45	-0.57	-0.28	-0.92	1

Algorithm 8.1. Model Tree for influent COD of Gouda

```
%% Model M5 generated with Weka using cross validation and P3,P2,P1,P and Q
% Correlation coefficient                      0.6814
% Mean absolute error                          50.7326
% Root mean squared error                      66.5986
% Relative absolute error                      70.8082 %
% Root relative squared error                  73.012  %
% Total Number of Instances                    393
load ('Q.mat');
n=length(Q);
COD=size(1:n,1);
% P3=P(:,1);
% P2=P(:,2);
% P1=P(:,3);
% P0=P(:,4);

for i=1:n
    if   Q(i,1) > 20845
        COD(i,1) =  - 4.0652 * P3(i,1)   ...
                    - 4.2365 * P2(i,1)   ...
                    + 0.1104 * P1(i,1)   ...
                    - 0.0031 * Q(i,1)    ...
                    + 484.5771;

    elseif  P3(i,1) <= 0.52 && Q(i,1) > 16625
        COD(i,1) =  168.2658 * P3(i,1)   ...
                    - 0.3437 * P2(i,1)   ...
                    + 0.6512 * P1(i,1)   ...
                    - 0.0014 * Q(i,1)    ...
                    + 486.6966;

    elseif P3(i,1) > 0.42
        COD(i,1) =  - 5.4033 * P3(i,1)   ...
                    - 0.7847 * P2(i,1)   ...
                    - 12.0484 * P1(i,1)  ...
                    + 4.2419 * P0(i,1)   ...
                    - 0.0021 * Q(i,1)    ...
                    + 481.3062;

    else
        COD(i,1) =  9.9928 * P1(i,1)   ...
                    + 498.9096;
    end
end
```

Table 8.12 Indicators of goodness of fit for COD models

Indicator of Goodness of Fit	Power Fucn	M5 tree
mean squared error (mse)	5163.21	3652.08
normalised mean squared error (nmse)	0.62	0.44
root mean squared error (rmse)	71.86	60.43
normalised root mean squared error (nrmse)	0.79	0.66
mean absolute error (mae)	54.90	46.42
mean absolute relative error (mare)	0.14	0.12
coefficient of correlation (r)	0.62	0.75
coefficient of determination (r-squared)	0.38	0.56
coefficient of efficiency (e)	0.38	0.56
maximum absolute error	241.67	215.37
maximum absolute relative error	0.93	0.61

Table 8.13 Correlation between variables for BOD model tree

	P3	P2	P1	P	Q	COD	BOD
P3	1	0.27	0.01	0.04	0.22	-0.48	-0.43
P2	0.27	1	0.15	0.06	0.34	-0.39	-0.34
P1	0.01	0.15	1	0.14	0.78	-0.32	-0.27
P	0.04	0.06	0.14	1	0.20	-0.09	-0.09
Q	0.22	0.34	0.78	0.20	1	-0.53	-0.50
COD	-0.48	-0.39	-0.32	-0.09	-0.53	1	0.78
BOD	-0.43	-0.34	-0.27	-0.09	-0.50	0.78	1

Algorithm 8.2. Model Tree for Influent BOD of Gouda

```
% Scheme:          weka.classifiers.rules.M5Rules -M 4.0
% === Cross-validation ===
% Correlation coefficient              0.7598
% Mean absolute error                 17.3759
% Root mean squared error             22.8855
% Relative absolute error             58.9869 %
% Root relative squared error         65.0369 %
% Total Number of Instances           145
%
% M5 pruned model rules
n=length(X);
BODm5=size(1,n);
for i=1:n
    if  COD(i,1) > 405
        BODm5(i,1) =   0.1226 .* P1(i,1)...
             - 0.0015 .* Qin(i,1)...
             + 0.1549 .* COD(i,1)...
             + 116.2049;% [77/64.187%]
    else
        BODm5(i,1) =   0.287 .* COD(i,1)...
             + 14.4219;% [68/75.642%]
    end
end
```

Table 8.14 Indicators of goodness of fit for BOD5 models

Indicators of Goodness of Fit	Lineal Func	M5 tree
mean squared error (mse)	489.74	423.29
normalised mean squared error (nmse)	0.40	0.34
root mean squared error (rmse)	22.13	20.57
normalised root mean squared error (nrmse)	0.63	0.59
mean absolute error (mae)	16.82	15.41
mean absolute relative error (mare)	0.12	0.11
coefficient of correlation (r)	0.78	0.81
coefficient of determination (r-squared)	0.60	0.66
coefficient of efficiency (e)	0.60	0.66
maximum absolute error	66.65	70.78
maximum absolute relative error	0.51	0.58

Table 8.15 Correlation between variables for TKN model tree

	P3	P2	P1	P0	Qin	TKN
P3	1	0.35	0.13	0.06	0.32	-0.48
P2	0.35	1	0.22	0.16	0.48	-0.56
P1	0.13	0.22	1	0.18	0.71	-0.52
P0	0.06	0.16	0.18	1	0.25	-0.19
Qin	0.32	0.48	0.71	0.25	1	-0.78
TKN	-0.48	-0.56	-0.52	-0.19	-0.78	1

Algorithm 8.3. Model Tree for influent TKN of Gouda

```
% Scheme:         weka.classifiers.rules.M5Rules -M 4.0
% Relation:       GoudaBOD
% Instances:      400
% === Classifier model (full training set) ===
% % === Cross-validation ===
% === Summary ===
%
% Correlation coefficient                    0.8488
% Mean absolute error                        3.7337
% Root mean squared error                    4.7629
% Relative absolute error                    50.7721 %
% Root relative squared error                52.7121 %
% Total Number of Instances              400
% M5 pruned model rules
% (using smoothed linear models) :
% Number of Rules : 2
n=length(X);
TKNm5=size(1,n);
for i=1:n
     if  Qin(i,1) <= 24575
        TKNm5(i,1) =    -0.5981 * P3(i,1)...
             - 0.3892 * P2(i,1)...
             - 0.4631 * P1(i,1)...
             + 0.1532 * P0(i,1)...
             - 0.001 * Qin(i,1)...
             + 68.2996;% [301/52.045%]
     else
         TKNm5(i,1)=      -0.1937 * P3(i,1)...
             - 0.3301 * P2(i,1)...
             + 0.147 * P0(i,1)...
             - 0.0004 * Qin(i,1)...
             + 49.3802;% [99/63.177%]
     end
end
```

Table 8.16 Indicators of goodness of fit for TKN models

Indicators of Goodness of Fit	Power Fucn	M5 tree
mean squared error (mse)	28.98	21.02
normalised mean squared error (nmse)	0.36	0.26
root mean squared error (rmse)	5.38	4.58
normalised root mean squared error (nrmse)	0.60	0.51
mean absolute error (mae)	4.14	3.52
mean absolute relative error (mare)	0.10	0.09
coefficient of correlation (r)	0.80	0.86
coefficient of determination (r-squared)	0.65	0.75
coefficient of efficiency (e)	0.64	0.74
maximum absolute error	25.50	18.92
maximum absolute relative error	0.98	0.68

Table 8.17 Correlation between variables for Ptot-P model tree

	P3	P2	P1	P0	Qin	Ptot
P3	1	0.35	0.17	0.06	0.38	-0.43
P2	0.35	1	0.27	0.16	0.53	-0.52
P1	0.17	0.27	1	0.22	0.67	-0.48
P0	0.06	0.16	0.22	1	0.27	-0.14
Qin	0.38	0.53	0.67	0.27	1	-0.74
Ptot	-0.43	-0.52	-0.48	-0.14	-0.74	1

Algorithm 8.4. Model Tree for Influent Ptot-P of Gouda

```
% === Classifier model (full training set) ===
% M5 pruned model rules
% (using smoothed linear models) :
% Number of Rules : 1
% === Cross-validation ===
% Correlation coefficient              0.758
% Mean absolute error                  0.7469
% Root mean squared error              0.9433
% Relative absolute error             64.1985 %
% Root relative squared error         65.0361 %
% Total Number of Instances            391
i=1;
PtotM5(:,i) = -0.0451 .* P3(:,i)...
            - 0.0448 .* P2(:,i)...
            + 0.0239 .* P0(:,i)...
            - 0.0001 .* Qin(:,i)...
            + 9.991;% [391/64.184%]
```

Table 8.18 Indicators of goodness of fit for Ptot-P models

Indicators of Goodness of Fit	Power Func	M5 tree
mean squared error (mse)	0.95	1.25
normalised mean squared error (nmse)	0.45	0.60
root mean squared error (rmse)	0.98	1.12
normalised root mean squared error (nrmse)	0.67	0.77
mean absolute error (mae)	0.77	0.92
mean absolute relative error (mare)	0.11	0.15
coefficient of correlation (r)	0.74	0.77
coefficient of determination (r-squared)	0.55	0.59
coefficient of efficiency (e)	0.54	0.40
maximum absolute error	3.14	2.90
maximum absolute relative error	0.45	0.67

Table 8.19 Mass Balances used to check the information measured in 2004

Balance over the treatment plant $Qef = Qin + Qsan + Qfil - Qex$ $Qef = Qin - Qex$ assumed to make the balance

Checking Qef and Qin with P balance	$Qef * Ptotef = Qin*Ptotin - Qex*Ptotex$	
P-tot influent	142.4 kg Ptot/d	calculated as: Ptot [kg P / m3] * Qin [m3/d]
P-tot effluent	10.1 kg Ptot/d	calculated as: Ptot [kg P / m3] * (Qin - Qex)[m3/d]
P-tot excess	131.4 kg Ptot/d	calculated as: Ptot [kg P / kg SS] * MLSS [kg SS / m3] * Qex [m3/d]
Balancing P tot	$Ptotef = Ptotin - Ptotex + Perror$	
error of Ptot	0.9 kg Ptot/d	
Error of Ptot was minimized by changing	$Qex =$ 929 m3/d	

Another way of checking the Qex is by matching the load of solids produced per day with the Qex and MLSS

Sludge Production	4063.0 kg sludge/d	from summary of data from 2004
MLSS	4.4 kg ss/m3	
SludgeProduction = Qex * MLSS		
$Qex =$	925.2 m3/d	

Balance base on Suspended Solids   Used to estimate the concentration of SS in the retourn sludge Gr

SS in the activated sludge		$Ga=((Qr*Gr) + (Qin * Gin))/(Qin + Qr)$	
SS in the retourn sludge		Grs	
	Ga	4.4 kg/m3 MLSS	Average measured in activated sludge (2004)
	Qr	26589 m3/d	assuming 1*Qin as retourn sludge
	Qin	23237 m3/d	
	Gin	0.10 kg SS/m3	Average value measured in influent flow(2004)
	Gr	8.1 kg/m3 MLSS	value estimated

The value estimated for Gr coincide with the assuptions made during the design =>

De volgende uitgangspunten zijn gehanteerd:
- slibgehalte actief-slibtanks = 3,48-4,00 kg/m³
- retourslibgehalte DWA = 8 kg/m³
- retourslibgehalte RWA = 10 kg/m³
- slibindex = 150 ml/g

Mass Balance for Total Nitrogen

Total Nitrogen Load in the Influent Ni	= Ne + Ns +Nd		
Total Nitrogen Load in the Effluent Ne		[kg/d]	
Total Nitrogen Load in the Excess Sludge Ns		[kg/d]	
Total Denitrified Nitrogen Load Nd		[kg/d]	
	Ni	929	Average value measured in influent flow(2004)
	Ne	170	Average value measured in effluent flow(2004)
	Ns + Nd	759	
	Ns	0.1 kg N/d	Calculated from the Average value of N per d.s.
	Nd	758.9 kg N/d	

Balance for Nitrification Nki = Nke+ Ns + Nn

Total Kjeldahl Nitrogen Load in the Inffluent Nki		
Total Kjeldahl Nitrogen Load in Effluent Nke		
Total Nitrogen Load in the Excess Sludge Ns		
Total Nitrified Nitrogen Load Nn		
	Nki	919
	Nke	65
	Ns + Nn	854
	Nn	854 kg N/d

Total Nitrified Nitrogen Load Nn = Nki - Nke - Nks		
Total Denitrified Nitrogen Load Nd = Nn + Nno3i - Nno3e - Nno3s		
	Nno3i	10.4 kg N-no3/d
	Nno3e	121.2 kg N-no3/d

Table 8.19 Mass Balances used to check the information measured in 2004 (Cont).

COD Balance to calculate the Oxygen Uptake in the treatment plant

COD Load in the Influent CODin = CODe + OUR + (Nd * 2.86) + (Qex * Gs.org * 1.42) - (4.56 * Nn)		
COD Load in the Effluent CODe		[kg O2/d]
Oxygen Uptake Rate OUR		[kg O2/d]
Volatile Suspended Soilds in Excess Sludge Gs.org		[kg VSS/m3]
TSS in excess sludge (assumed MLSS)	4.4	kg SS/m3
ration of kg COD / kg VSS	1.42	
Oxygen Uptake during Nitrification	4.56	
CDOin	8711	[kg O2/d]
CODe	770	[kg O2/d]
Qex	929	m3/d
Gs.org	3.7	[kg VSS/m3]
Nd	759	kg N/d
Nn	854	kg N/d
OUR	4740	[kg O2/d]
Verbruik Beluchting	3140	kWh/d
Aeration Efficiency of the Plant	1.51	kgO2/kWh

Calculation of SRT from the P balance

Total Reactor Volume	29,596 m3	
P-tot in activated sludge (Ptot-r)	142.73 g/m3	
PO4-eff	0.26 g/m3	
P_TSS_r	142.47 g/m3	
Qin	23237 m3/d	
Ptot-in	6.07 g/m3	
SRT =	31.3 d	SRT calculated with out excess measurements
Qex =	929 m3/d	
Qeff =	25660 m3/d	
P-TSS_ex	142.5 g/m3	
P_tot_eff =	0.39 g/m3	
P_TSS_eff	0.14 mgP/l	
SRT =	31.0 d	SRT calculated with the balance excess sludge flow
TSS in activated sludge reactor TSS_r	4391.7 g/m3	
TSS_eff	4 g/m3	
TSS_ex	4391.7 g/m3	
SRT =	31.1 d	SRT calculated with the TSS
SRT = Vol/Qex	31.9 d	

211

Table 8.20 Gouda Wastewater Characterisation

Wastewater Characterisation for ASM2d	Equation or conversion factor		Average / Calculated	Notes
COD fractionation				
Non filtered total COD influent	CODinf,tot =	Sa + Sf + Si + Xs + Xi + Xh + Xaut + Xpao + Xpha		
Assumtion :	Xh, Xpha = 0,	Xaut, Xpao = 0.1 - 1		
Then	CODinf,tot =	Sa + Sf + Si + Xs + Xi	371.5	
COD influent after membrane filtration (0.1um)	CODinf,sol			
	CODinf,tot =	CODinf,sol + CODinf,part		
Assumtion from Roeleveld and van Loosdrecht (2002)	CODinf,sol/CODinf,to		0.27	
	CODinf,sol =	Sa + Sf + Si	101.0	
	CODinf,part =	Xs + Xi	270.5	
BOD influent non filtered +ATU + inoculum of effluent wwtp	BOD5,inf =		153.6	
Assumption based on measures by RandL(2002)	Kbod =		0.38	
BOD total calculated from BOD5 inf	BODtot,inf =		5	180.6
Biodegradable COD calculated from BODtot with fbod	fbod=		0.05	
	BCOD =	Sa + Sf + Xs	190.1	
COD effluent	CODeff,tot		32.8	
Assumption that CODeff,sol/CODeff,tot			0.9	
COD effluent after membrane filtration (0.1um)	CODeff,sol		29.6	
	Si =	0.9 * CODeff,sol (low loaded wwtp)	26.6	
	Ss =	CODinf,sol - Si	74.4	
CODvfa measured by gas chromatography or titration	Sa =	CODvfa	11.2	
Asumming that Sa is a fraction of the Ss			0.15	fbod =
	Sf =	Ss - Sa	63.2	Xs/(Xi+Xs)
	Xs =	BCOD - Ss	115.7	Xs/Xi
	Xi =	CODinf,tot - Si - Ss - Xs	154.8	Si/Sf
				Xi/Xs
Nitrogen fractionation				
				Typical Ranges
Conversion factors for Nitrogen				gN/gCOD
	i_{NSi}	0.02 gN/gCOD		0.01 - 0.02
	i_{NSA}	0 gN/gCOD		0
	i_{NSF}	0.03 gN/gCOD		0.02 - 0.04
	i_{NXI}	0.03 gN/gCOD		0.01 - 0.06
	i_{NXS}	0.04 gN/gCOD		0.02 - 0.06

$$S_{NH4} = NH4\text{-}N_{inf}$$
$$N\text{-}Kj_{inf,tot} = S_{NH4} + (Sf*I_{NSF}) + (Si*I_{NSi}) + (Xi*I_{NXi}) + (Xs*I_{NXs})$$

	N-Kjinf,tot =		39.2	
	Sf*INSF =		1.9	
	Si*INSi =		0.5	
	Xi*INXi =		4.6	
	Xs*INXs =		4.6	SNH4/N-Kjinf
	SNH4 =		27.5	
	SNO3 =		0.45	0.45
Phosphorus fractionation				
				Typical Ranges
Conversion factors for Phosphorus				gP/gCOD
	i_{PSi}	0 gP/gCOD		gP/gCOD
	i_{PSA}	0 gP/gCOD		0.002 - 0.008
	i_{PSF}	0.01 gP/gCOD		0
	i_{PXI}	0.025 gP/gCOD		0.01 - 0.015
	i_{PXS}	0.01 gP/gCOD		0.005 - 0.01
				0.01 - 0.015

$$S_{PO4} = PO4\text{-}P_{inf}$$
$$P_{inf,tot} = S_{PO4} + (Sf*I_{PSF}) + (Si*I_{PSi}) + (Xi*I_{PXi}) + (Xs*I_{PXs})$$

	Pinf,tot =		6.1	
	Sf*IPSF =		0.6	
	Si*IPSi =		0.0	
	Xi*IPXi =		3.9	
	Xs*IPXs =		1.2	
	SPO4 =		0.41	

Suspended Solids Fractionation
Important Issue: in STOAT, TSS has to include the XS; if

not the nVSS will be calculated NEGATIVE	TSS= sum of particulated components Xi*0.75*Xs*0.75+Xh*0.9			
Total suspended solids concentration	**TSS**		203.7	mg SS/L
Volatile suspended solids concentration	**VSS**	VSS = 0.85*TSS	173.2	mg/L
Non-biodegradable Volatile Suspended Solids in the influent	**nbVSS**	nbVSS = 0.33*VSS	57.2	mg/L
Non Volatile suspended solids	**nVSS**	TSS - VSS	30.6	
Nota for the model Biomass at the influent was assumed as follow				
Heterotrophic biomass	**Xh**	1		mg COD/l
Autotrophic, nitrifying biomass	**Xaut**	0.1		mg COD/l
Phosphorus accumulating organisms	**Xpao**	0.1		mg COD/l

212

Table 8.21 Parameters and coefficients used in the model of the Gouda WwTP

Item	Stoichiometric Parameters	Default STOAT	ID	ASM2d	Changed from STOAT
1	Fractional hydrolysis rate anoxic conditions (-)	0.6	ηNO3	0.6	
2	Fractional hydrolysis rate anaerobic conditions (-)	**0.1**	ηfe	**0.4**	**0.4**
3	Fractional anoxic growth rate (heterotrophs) (-)	0.8	ηNO3	0.8	
4	Fractional anoxic growth rate (PAO) (-)	0.6	ηNO3	0.6	
5	Heterotroph yield (mg COD/mg COD)	0.63	YH	0.625	
6	Yield coefficient on PAO (mg COD/mg COD)	0.63	YPAO	0.625	
7	Autotroph yield (mg COD/mg N)	0.24	YA	0.24	
8	Inert COD generated in heterotroph lysis (mg COD/mg COD)	0.1	fXI	0.1	
9	Inert COD generated in PAO lysis (mg COD/mg COD)	0.1	fXI	0.1	
10	Inert COD generated in autotroph lysis (mg COD/mg COD)	0.1	fXI	0.1	
11	Fraction of inert COD in particulate substrate (mg COD/mg COD)	0	fSI	0	
12	PHA requirement for PolyP storage (mg COD/mg P)	0.2	YPHA	0.2	
13	PolyP requirement for PHA storage (mg P/mg COD)	0.4	YPO4	0.4	
14	N content of biomass (mg N/mg COD)	0.07	iNBM	0.07	
15	P content of biomass (mg P/mg COD)	0.02	iPBM	0.02	
16	N content of inert soluble COD (mg N/mg COD)	0.01	iNSI	0.01	0.02
17	N content of soluble degradable COD (mg N/mg COD)	0.03	iNSF	0.03	
18	N content of inert particulate COD (mg N/mg COD)	**0.03**	iNXI	**0.02**	**0.03**
19	N content of particulate degradable COD (mg N/mg COD)	0.04	iNXS	0.04	
20	P content of inert soluble COD (mg P/mg COD)	0	iPSI	0	
21	P content of soluble degradable COD (mg P/mg COD)	0.01	iPSF	0.01	
22	P content of inert particulate COD (mg P/mg COD)	**0.01**	iPXI	**0.025**	**0.025**
23	P content of degradable particulate COD (mg P/mg COD)	0.01	iPXS	0.01	
24	TSS to inert particulate COD ratio (mg/mg COD)	0.75	iTSSXI	0.75	
25	TSS to degradable COD ratio (mg/mg COD)	0.75	iTSSXS	0.75	
26	TSS to biomass COD ratio (mg/mg COD)	0.9	iTSSBM	0.9	
27	TSS to metal salt ratio (mg/mg)	3.45			
28	TSS to metal phosphate ratio (mg/mg)	4.87			

	Kinetic Parameters	Default STOAT 15oC	ID	ASM2d 20oC	Changed from STOAT
29	Heterotroph growth rate @ 15°C (1/h)	0.176776	µH	0.25	
30	Heterotroph temperature coefficient (1/°C)	0.069314	a	1.07	
31	PAO growth rate @ 15°C (1/h)	0.034105	µPAO	0.0417	
32	PAO temperature coefficient (1/°C)	0.040047	a	1.04	
33	Autotroph growth rate @ 15°C (1/h)	0.02465	µAUT	0.0417	
34	Autotroph temperature coefficient (1/°C)	0.104982	a	1.12	
35	Heterotroph death rate @ 15°C (1/h)	0.011785	bH	0.0167	
36	Heterotroph temperature coefficient (1/°C)	0.06934	a	1.07	
37	PAO death rate @ 15°C (1/h)	0.005892	bPAO	0.0083	
38	PAO temperature coefficient (1/°C)	0.06934	a		
39	Autotroph death rate @ 15°C (1/h)	0.003608	bAUT	0.0063	
40	Autotroph temperature coefficient (1/°C)	0.109861	a		
41	Liberation rate of polyP @ 15°C (1/h)	0.005892	bPP	0.0083	
42	PolyP liberation temperature coefficient (1/°C)	0.069314	a		
43	Liberation rate of PHA @ 15°C (1/h)	0.005892	bPHA	0.0083	
44	PHA liberation temperature coefficient (1/°C)	0.069314	a		
45	Hydrolysis rate @ 15°C (1/h)	0.102062	Kh	0.1250	
46	Hydrolysis temperature coefficient (1/°C)	0.040546	a		
47	Hydrolysis half-rate constant @ 15°C	0.007216	Kx	0.0042	
48	Half-rate constant temperature coefficient (1/°C)	**-0.10986**	a		**0.10986**
49	Fermentation rate @ 15°C (1/h)	0.088388	qfe	0.125	
50	Fermentation temperature constant (1/°C)	0.069314	a		
51	PHA uptake rate @ 15°C (1/h)	0.102062	qPHA	0.1250	
52	PHA uptake temperature coefficient (1/°C)	0.040546	a		
53	PolyP uptake rate @ 15°C (1/h)	0.051031	qPP	0.0625	
54	PolyP uptake temperature coefficient (1/°C)	0.040546			
55	Precipitation rate @ 15°C (1/h)	0.041666	KPRE	0.0417	
56	Precipitation temperature coefficient (1/°C)	0			
57	Dissolution rate @ 15°C (1/h)	0.025	KRED	0.025	
58	Dissolution temperature coefficient (1/°C)	0			

	Switching Coefficients	Default STOAT	ID	ASM2d	Changed from STOAT
59	O2 half-rate constant (hydrolysis) (mg O2/l)	0.2	KO2	0.2	
60	O2 half-rate constant (heterotrophs) (mg O2/l)	0.2	KO2	0.2	
61	O2 half-rate constant (PAO) (mg O2/l)	0.2	KO2	0.2	
62	O2 half-rate coefficient (autotrophs) (mg O2/l)	0.5	KO2	0.5	
63	NO3 half-rate constant (hydrolysis) (mg N/l)	0.5	KNO3	0.5	
64	NO3 half-rate constant (heterotrophs) (mgN/l)	0.5	KNO3	0.5	
65	NO3 half-rate constant (PAO) (mg/l)	No exist	KNO3	0.5	0.2
66	NH4 half-rate constant (heterotrophs) (mg N/l)	0.05	KNH4	0.05	
67	NH4 half-rate constant (PAO) (mg N/l)	0.05	KNH4	0.05	
68	NH4 half-rate coefficient (autotrophs) (mg N/l)	1	KNH4	1	0.8
69	SCOD half-rate constant (heterotrophs) (mg COD/l)	4	KF	4	
70	SCOD half-rate constant (fermentation) (mg COD/l)	**20**	Kfe	**4**	**4**
71	VFA half-rate constant (heterotrophs) (mg COD/l)	4	KA	4	
72	VFA half-rate constant (PAO) (mg COD/l)	4	KA	4	
73	P half-rate constant (heterotrophs) (mg P/l)	0.01	KP	0.01	
74	P half-rate constant PolyP uptake (mg P/l)	0.2	KPS	0.2	
75	P half-rate constant (PAO) (mg P/l)	0.01	KP	0.01	
76	P half-rate coefficient (autotrophs) (mg P/l)	0.01	KP	0.01	
77	PolyP half-rate constant (mg P/l)	0.01	KPP	0.01	
78	Maximum P storage (mg P/mg COD)	0.34	KMAX	0.34	
79	Inhibition coefficient for PolyP (mg P/l)	0.02	KIPP	0.02	
80	Saturation coefficient for PHA (mg COD/l)	0.01	KPHA	0.01	

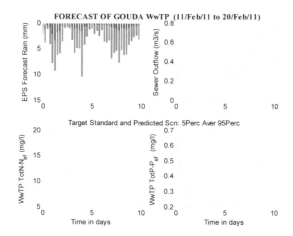

Figure 8.6. Forecast of the disturbance at the treatment plant

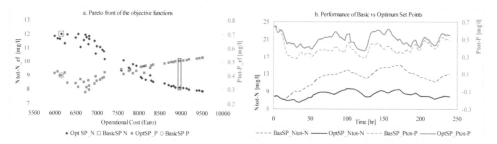

Figure 8.7 Solutions of the optimization of the set points. a. Pareto sets b. Performance of selected solution compared with the basic set points.

Table 8.22 Comparison of the objective functions for the basic and optimum set points

	Set Points			Optimization Objectives		
	Qir	DO4	DO5	Ntot-N	Ptot-P	Oper Cost
	[m3/h]	[mg/l]	[mg/l]	[mg/l]	[mg/l]	[Euro]
BasicSP	1161	1	0.5	12.0	0.40	6146
OptSP	2158	3.9	1.9	8.2	0.50	8759

9 References

Achleitner, S.; Moderl, M.; Rauch, W.,(2007). "City drain (c) - an open source approach for simulation of integrated urban drainage systems", *Environmental Modelling & Software*, Vol.22 (8), p.p. 1184.

Afshar, M. H.,(2006). "Application of a max-min ant system to joint layout and size optimization of pipe networks", *Engineering Optimization*, Vol.38 (3), p.p. 299

Afshar, M. H.,(2008). "Rebirthing particle swarm optimization algorithm: Application to storm water network design", *Canadian Journal of Civil Engineering*, Vol.35 (10), p.p. 1120.

Afshar, M. H.; Afshar, A.; MariÃ±o, M. A.; Darbandi, A. A. S.,(2006). "Hydrograph-based storm sewer design optimization by genetic algorithm", *Canadian Journal of Civil Engineering*, Vol.33 (3), p.p. 319.

Ambrose, R. B.; Wool, T. A.,(2009). "Wasp7 stream transport - model theory and user's guide. Supplement to water quality analysis simulation program (wasp) user documentation."U.S. Environmental Protection Agency Office of Research and Development National Expsoure Research Laboratory Ecosystems Research Division. Athens, G.,

Ambrose, R. B.; Wool, T. A.; Martin, J. L., (1993). "The water quality analysis simulation program, wasp5; part a: Model documentation", In U.S. Environmental Protection Agency, C. f. E. A. M. Ed. Athens, GA.

Andel, S. J. v., (2009)."Anticipatory water management. Using ensemble weather forecasts for critical events". Hydroinformatics and Knowledge Management, Unesco-IHE Institute for Water Education and Delft University of Technology. Delft.

Andel, S. J. v.; Price, R. K.; Lobbrecht, A.; Kruiningen, F. v.; Mureau, R.,(2008). "Ensamble precipitation and water-level forecast for anticipatory water-system control", *Atmosferic science letters*, Vol.9 (2), p.p. 57~60.

Anne, A.; Johannes, B.; Dimo, B.; Eckart, Z., (2009). "Theory of the hypervolume indicator: Optimal µ-distributions and the choice of the reference point", In *Proceedings of the tenth ACM SIGEVO workshop on Foundations of genetic algorithms*, ACM: Orlando, Florida, USA.

Arnbjerg-Nielsen, K.,(2006). "Significant climate change of extreme rainfall in denmark", *Water Sci Technol*, Vol.54 (6-7), p.p. 1.

Azarm, S.; Wu, J.,(2001). "Metrics for quality assessment of a multiobjective design optimization solution set", *Journal of Mechanical Design*, Vol.123 (1), p.p. 18.

Back, T.; Fogel, D.; Michalewicz, Z., (2000). "*Handbook of evolutionary computation*", IOP Publishing Ltd. and Oxford University Press Bristol, U.K.

Bader, J.; Zitzler, E., (2008). "Hype: An algorithm for fast hypervolume-based many-objective optimization", In Laboratory, C. E. a. N. Ed. Computer Engineering and Networks Laboratory: Zurich, p 25.

Barreto, W.; Vojinovic, Z.; Price, R.; Solomatine, D., (2008). "Multi-tier modelling of urban drainage systems: Muti-objective optimization and parallel computing", In *11th International Conference on Urban Drainage*, Edinburg.

Barreto, W. J.; Price, R. K.; Solomatine, D. P.; Vojinovic, Z., (2006). "Approaches to multi-objective multi-tier optimization in urban drainage planning", In *7th International Conference on Hydroinformatics HIC 2006*, Nice, France.

Benedetti, L.; Krebs, P.; Vanrolleghem, P.,(2004b). "Cost-effective development of urban wastewater systems for water framework directive compliance–the cd4wc eu project", p.p.

Bokhorst, J.; Lobbrecht, A.,(2005). "Neerslag-kansverwachting voor het waterbeheer. (probability precipitatio forecast for water management)", *H2O*, Vol.Vol 13 p.p. 26.

Branke, J. u.; Schmeck, H.; Deb, K.; Reddy.S, M., (2004). "Parallelizing multi-objective evolutionary algorithms: Cone separation", In KanGAL Ed.

Brdys, M. A.; Grochowski, M.; Gminski, T.; Konarczak, K.; Drewa, M.,(2008). "Hierarchical predictive control of integrated wastewater treatment systems", *Control Engineering Practice*, Vol.16 (6), p.p. 751~767.

Brown, L. C.; Barnwell, T. O., (1987). "*The enhanced stream water quality models: Qual2e and qual2e-uncas: Documentation and user model*", U.S. Environmental Protection Agency, EPA/600/3-87/007: Athens

Burger, G.; Fach, S.; Kinzel, H.; Rauch, W., (2009). "Parallel computing in integrated urban drainage simulations", In *8th International Conference on Urban Drainage Modelling*, Tokyo, Japan.

Burgess, E. H.; R.Magro, W.; Clement, M. A.; Moore, C. I.; Smullen, J. T., (1999). "Parallel processing enhancement to swmm/extran", In *Applied modeling of urban water systems, monograph 8 in the series, proceedings of the conference on stormwater and urban water systems modeling*, James, W. Ed. CHI: Toronto, Ontario, , p 45~60.

Burian, S. J.; Nix, S. J.; Durrans, S. R.; Pitt, R. E.; Fan, C.; Field, R.,(1999). "Historical development of wet-weather flow management", *Journal of Water Resources Planning and Management*, Vol.125 (1), p.p. 3.

Burn, D. H.; Yullanti, J. S.,(2001). "Waste-load allocation using genetic algorithms", *Journal of Water Resources Planning and Management*, Vol.127 (2), p.p. 121~129.

Butler, D.; Schütze, M.,(2005). "Integrating simulation models with a view to optimal control of urban wastewater systems", *Environmental Modelling & Software*, Vol.20 (1), p.p. 415~426.

Butler, D. E.; Davies, J., (2000). "*Urban drainage*", Spon Press (UK).

Cantu-Paz, E., (1999). "Implementing fast and flexible parallel genetic algorithms", In *Practical handbook of genetic algorithms*, Chambers, L. D. Ed. CRC Press: Vol. volume III, pp 65.

Cao, Y.,(2008). "Function v=hypervolume(p,r,n) version 1.0", MATLAB Central: http://www.mathworks.com/matlabcentral/fileexchange/19651-hypervolume-indicator.

Casares, J. J.; Rodriguez, J.,(1989). "Analysis and evaluation of a wastewater treatment plant model by stochastic optimization", *Applied Mathematical Modelling*, Vol.13 (7), p.p. 420.

Castelletti, A.; Pianosi, F.; Soncini-Sessa, R.; Antenucci, J. P.,(2010). "A multi-objective response surface approach for improved water quality planning in lakes and reservoirs", *Water Resources Research*, Vol.46, W06502 p.p. 16.

CEC, (1991). "Council of the european communities directive concerning urban wastewater treatment (91/271/eec)", In Official Journal of European Communities, L135: pp 40

CEC, (2000). "Council of the european communities. Directive 2000/60/ec of the european parliament and of the council of 23 october 2000 establishing a framework for community action in the field of water policy", In Official Journal of the European Communities, L327: pp 1

CEC, (2007). "Commission of european communities. Communication from the commission to the european parliament and the council. Towards sustainable water management in the european union - first stage in the implementation of the water framework directive 2000/60/ec –", In *[SEC(2007) 362] [SEC(2007) 363]*, Brussels, pp 1.

Cembrowicz, R. G.,(1994). "Evolution strategies and genetic algorithms in water supply and waste water systems design", *Transactions on Ecology and the Environment*, Vol.7 p.p.

Chapra, S. C., (1997). "*Surface water quality modeling* ", McGraw-Hill: Boston, p 844

Cho, J. H.; Seok Sung, K.; Ryong Ha, S.,(2004). "A river water quality management model for optimising regional wastewater treatment using a genetic algorithm", *Journal of Environmental Management*, Vol.73 (3), p.p. 229.

Clemens, F. H. L. R., (2001)."Hydrodynamic models in urban drainage: Application and calibration". Delft University of Technology. Delft.

Clifforde, I. T.; Tomicic, B.; Mark, O.,(1999). "Integrated wastewater management-a european vision for the future", *Proc. the Eighth International Conference on Urban Storm Drainage*, p.p. 168.

Coello Coello, C.; Lamont, G. B.; Veldhuizen, D. A. V., (2002). "*Evolutionary algorithms for solving multi-objective problems (genetic algorithms and evolutionary computation)*", Kluwer Academic/Plenum publishers.: New York.

Coello Coello, C.; Toscano Pulido, G.; Zitzler, E.; Thiele, L.; Deb, K.; Coello Coello, C.; Corne, D., (2001). "A micro-genetic algorithm for multiobjective optimization evolutionary multi-criterion optimization", In Springer Berlin / Heidelberg: Vol. 1993, pp 126.

Copp, J. B. Ed. (2002). "*The cost simulation benchmark: Description and simulator manual*", Office for Official Publications of the European Communities.: Luxembourg, Vol.

Corne, D. W.; Jerram, N. R.; Knowles, J. D.; Oates, M. J.; J, M., (2001). "Pesa-ii: Region-based selection in evolutionary multiobjective optimization", In *Genetic and Evolutionary Computation Conference (GECCO'2001)*, Morgan Kaufmann Publishers: p 283~290.

Creel, M.,(2005). "User-friendly parallel computations with econometric examples", *Computational Economics*, Vol.26 p.p. 107.

Creel, M.; Goffe, W. L.,(2008). "Multi-core cpus, clusters, and grid computing: A tutorial", *Computational Economics*, Vol.Volume 32 (4), p.p. 29.

CVC, (2009). "Curva de duracion de caudales para el río lili. Estacion pasoancho", In Corporacion autónoma regional del Valle del Cauca: Cali.

Cyclus NV,(2008). "Cyclus nv. Onze producten en diensten. Riolering." http://www.cyclusnv.nl/#pagina=1084:

DAGMA, (2006). "Resolución no. 376 de 2006 por medio de la cual se establecen los objetivos de calidad para los cuerpos de agua en el área urbana del municipio de cali para el período 2007 - 2016", In Departamento Administrativo de Gestión del Medio Ambiente – DAGMA Ed. pp 1

DAGMA; UNIVALLE, (2009). "Identificación de los responsables por vertimientos, captaciones de agua, riesgos y vulnerabilidad en el río lilí, la quebrada gualí y el zanjón del burro, en el área urbana de santiago de cali", In Cali.

Dajani, J. S.; Hasit, Y.,(1974). "Capital cost minimization of drainage networks", *Journal Environmental Engeniering. ASCE.*, Vol.100 (2), p.p. 325~337.

Deb, K.; Agrawal, S.; Pratap, A.; Meyarivan, T.,(2002). "A fast elitist non-dominated sorting genetic algorithm for multi-objective optimization: Nsga-ii ", *IEEE Transactions on Evolutionary Computation*, Vol.6 (2), p.p. 181 ~ 197.

Diogo, A. F.; Walters, G. A.; Ribeiro de Sousa, E.; Graveto, V. M.,(2000). "Three-dimensional optimization of urban drainage systems", *Computer-Aided Civil and Infrastructure Engineering*, Vol.15 (6), p.p. 409.

Diwekar, U. M., (2003). "*Introduction to applied optimization*", Kluwer Academic Norwell Mass.

EC,(2003). "Guidance document no 7. Monitoring under the water framework directive",Working Group 2.7 - Monitoring,European Communities (EC),

EEC,(2007). "Terms and definitions of the urban waste water treatment directive (91/271/eec)",UWWTD-REP Working Group,

Egea, J. A.; Vries, D.; Alonso, A. A.; Banga, J. R.,(2007). "Global optimization for integrated design and control of computationally expensive process models", *Ind. Eng. Chem. Res.*, Vol.46 p.p. 9148.

EMCALI, (2004). "Resolución no. 000147 de 19 de febrero de 2004", In Gerencia General de EMCALI E.I.C.E. E.S.P., Cali.

EMCALI; Hidro-Occidente SA, (2006). "Estudio de alternativas de dotación de los servicios públicos de acueducto, alcantarillado y complementarios de alcantarillado en la zona de expansión de la ciudad de cali denominada "Corredor cali-jamundí"." In Empresas Municipales de Cali: Cali.

EPA, (2006). "Real time control of urban drainage networks", In EPA's Office of Research and Development Ed. United State Environment Protection Agency: Washington, p 96.

Erbe, V.; Risholt, L. P.; Schilling, W.; Londong, J.,(2002). "Integrated modelling for analysis and optimisation of wastewater systems-the odenthal case", *Urban Water*, Vol.4 (1), p.p. 63.

Farmani, R.; Savic, D. A.; Walters, G. A., (2006). "A hybrid technique for optimization of branched urban water systems", In *7th Int. Conf. of Hydroinformatics*, Nice, France, Vol. 1, p 985~992.

Freni, G.; Maglionico, M.; Federico, V., (2003). "State of the art in urban drainage modelling", In *WP3 - Hydraulic performance*, Computer Aided REhabilitation of Sewer networks CARE-S Project.

Freni, G.; Mannina, G.; Viviani, G.,(2009). "Uncertainty assessment of an integrated urban drainage model", *Journal of Hydrology*, Vol.373 (3-4), p.p. 392~404.

Freundy, Y.; Seung, H. S.,(1997). "Selective sampling using the query by committee algorithm", *Machine Learning*, Vol.28 (133), p.p.

Fu, G.; Butler, D.; Khu, S.-T.,(2007). "Multiple objective optimal control of integrated urban wastewater systems", *Environmental Modelling & Software*, (23), p.p. 225 ~ 234.

Fu, G.; Butler, D.; Khu, S.-T.,(2008). "Multiple objective optimal control of integrated urban wastewater systems", *Environmental Modelling and Software*, Vol.23 p.p. 225.

Fu, G.; Khu, S.-T.; Butler, D., (2008). "Multiobjective optimisation of urban wastewater systems using parego: A comparison with nsga ii", In *11th International Conference on Urban Drainage,*, Edinburgh, Scotland, UK. .

FWR, (1994). "Urban pollution management manual. 1st edition, fr/cl 0002", In Foundation for Water Research.: Marlow, UK.

Gannon, D.; Reed, D., (2009). "Parallelism and the cloud", In *The fourth paradigm: Data-intensive scientific discovery*, Hey, T.; Tansley, S., *et al.* Eds.; Microsoft Research: Washingtong, pp 131

Gaspar-Cunhaa, A.; Vieira, A.,(2003). "A multi-objective evolutionary algorithm using neural networks to approximate fitness evaluations", *International Journal of Computers, Systems and Signals*, Vol.Vol 6, No.1 (No.1), p.p. Pag 18.

218

Gemeente Gouda, (2004). "Ontwerp gemeentelijk rioleringsplan gouda 2004 - 2008", In *Journal of Water Resources Planning and Management*, Gouda.

Glover, F.; Laguna, M., (1997). "*Tabu search*", Kluwer Academic Publishers: Boston, Massachusetts.

Goldberg, D. E., (1989). "*Genetic algorithms in search, optimization and machine learning.* " Addison-Wesley Publishing Company: Reading, Massachusetts.

González, L.; Peñaranda, L., (2004). "Factores que contribuyen en la extinción de la microcuenca del río lilí (cali - colombia). " In Organización Internacional de Universidades por el Desarrollo Sostenible y el Medio Ambiente: Nicaragua.

Govindarajan, L.; Kumar, S. K.; Karunanithi, T.,(2005). "Optimal design of wastewater treatment plant using adaptive simulated annealing", *Journal of Applied Sciences and Environmental Management*, Vol.9 (1), p.p. 107~113.

Grum, M.; Jorgensen, A. T.; Johansen, R. M.; Linde, J. J.,(2006). "The effect of climate change on urban drainage: An evaluation based on regional climate model simulation", *Water Sci Technol*, Vol.54 (6-7), p.p. 9.

Guo, Y.; Walters, G.; Khu, S.-T.; Keedwell, E.,(2007). "Optimal design of sewer networks using hybrid cellular automata and genetic algorithm", *Engineering Optimization*, Vol.39 (3), p.p. 345~364.

Guo, Y.; Walters, G.; Savic, D., (2008). "Optimal design of storm sewer networks: Past, present and future", In *Proceedings of the 11th International Conference on Urban Drainage*, Edinburgh, Scotland, UK.

Gupta, A.; Mehndiratta, S. L.; Khanna, P.,(1983). "Gravity wastewater collection systems optimization", *Journal of Environmental Engineering, ASCE*, Vol.109 (5), p.p. 1195~1209.

Gustafson, J. L., (1988). "Reevaluating amdahl's law", In Sandia National Laboratories.

Gutiérrez, G.; Vega, P., (2002). "Process synthesis applied to activated sludge processes: A framework with minlp optimisation models", In *15th IFAC Triennial World Congress*, Barcelona.

Harremoes, P., (2002). "Integrated urban drainage, status and perspectives", In *Interactions Between Sewers, Treatment Plants and Receiving Waters in Urban Areas (INTERURBA II)*, Saldanha, J. Ed. Water Science and Technology IWA Publishing: Lisbon - Portugal, Vol. 45, pp 1.

Harremoës, P., (2002). "Integrated urban drainage, status and perspectives", In *Interactions Between Sewers, Treatment Plants and Receiving Waters in Urban Areas (INTERURBA II)*, Saldanha, J. Ed. Water Science and Technology IWA Publishing: Lisbon - Portugal, Vol. 45, p 1~10.

Harremoës, P.; Rauch, W.,(1996). "Integrated design and analysis of drainage systems, including sewers, treatment plant and receiving waters", *Journal of Hydraulic Research*, Vol.34 (6), p.p. 815.

Harremoës, P.; Rauch, W.,(1999). "Optimal design and real time control of the integrated urban run-off system", *Hydrobiologia*, Vol.410 p.p. 177.

Hassan, R.; Scholes, R.; Ash, N. Eds., (2005). ""*Ecosystems and human well-being: Current state and trends, volume 1". Millenium ecosystem assessment. Findings of the condition and trends working group*", Island Press: Vol. Vol 1.

Heaney, J. P.; Sample, D.; Wright, L.,(2002). "Costs of urban stormwater control",U.S. Environmental Protection Agency,EPA.,

Henze, M.; Gujer, W.; Mino, T.; Matsuo, T.; Wentzel, M. C.; Marais, G. v. R.; van Loosdrecht, M. C. M.,(1999). "Activated sludge model no. 2d." *Water Science and Technology*, Vol.39 (1), p.p. 165~182.

219

T - #0144 - 160425 - C0 - 240/170/14 - PB - 9781138000025 - Gloss Lamination